GUIDE FOR MUNICIPAL WET WEATHER STRATEGIES

WEF Special Publication

Second Edition

2013

Water Environment Federation
601 Wythe Street
Alexandria, VA 22314-1994 USA
http://www.wef.org

Guide for Municipal Wet Weather Strategies

ISBN 978-1-57278-278-5

About WEF

Founded in 1928, the Water Environment Federation (WEF) is a not-for-profit technical and educational organization of 36,000 individual members and 75 affiliated Member Associations representing water quality professionals around the world. WEF members, Member Associations, and staff proudly work to achieve our mission to provide bold leadership, champion innovation, connect water professionals, and leverage knowledge to support clean and safe water worldwide. To learn more, visit www.wef.org.

For information on membership, publications, and conferences, contact

Water Environment Federation
601 Wythe Street
Alexandria, VA 22314-1994 USA
(703) 684-2400
http://www.wef.org

Prepared by the **Guide for Municipal Wet Weather Strategies** Task Force of the **Water Environment Federation**

Authors
Daniel W. Ott, P.E., *Co-Chair*
Reggie Rowe, P.E., *Co-Chair*
Nancy J. Wheatley, *Co-Chair*

Jacqueline Kepke, P.E.
Andy Lukas, P.E.
Jane Madden, P.E.
S. Wayne Miles, P.E., BCEE
Julian Sandino, Ph.D., BCEE., P.E.
Nancy Schultz, P.E., D.WRE
Mike Stickley, P.E.
Gary Swanson, P.E.
Christopher W. Tabor, P.E.
Brandon C. Vatter, P.E.

Technical Reviewers
Laith Alfaqih, Ph.D., P.E.
Rich Atoulikian, PMP, BCEE, P.E.
Scott E. Belz
Geri Blakey
Michael L. Carr, P.E.
Chein Chi Chang, Ph.D., P.E.
Caitlin Feehan, P.E., LEED AP
Richard Finger
James D. Fitzpatrick, P.E.
Jeffery Frey, P.E.
Thomas M. Grisa, P.E., F.ASCE
Eric M. Harold, P.E., BCEE
Gregory R. Heath, P.E.
Deborah A.S. Hoag, P.E.
Carol L. Hufnagel, P.E.
Lawrence P. Jaworski, P.E., BCEE
Sharon A. Jean-Baptiste, P.E.
Brandon J. Koltz
Xiangfei Li, P.Eng., Ph.D.
Jim Liubicich, P.E.
Helen (Haiyi) Lu, P.E.
Waldo Margheim, P.E.
Robert L. Matthews, P.E., BCEE
John L. Murphy, P.E.
Pat Nelson, P.E.
Richard (Rick) E. Nelson, P.E.
Bipin Pathak, Ph.D., P.E.
David Powell, Ph.D., P.E.
Neil D. Raymond
Christopher Reichle, P.E.
Michael Sevener, P.E., BCEE
Saša Tomi, Ph.D., P.E., BCEE
Barry Tonning
Michel Wanna
David Winters, P.E.
Paula Zeller, CA-WWTPO Grade V

Under the direction of the **Stormwater Subcommittee** of the **Technical Practice Committee**

2013

Water Environment Federation
601 Wythe Street
Alexandria, VA 22314-1994 USA
http://www.wef.org

Special Publications of the Water Environment Federation

The WEF Technical Practice Committee (formerly the Committee on Sewage and Industrial Wastes Practice of the Federation of Sewage and Industrial Wastes Associations) was created by the Federation Board of Control on October 11, 1941. The primary function of the Committee is to originate and produce, through appropriate subcommittees, special publications dealing with technical aspects of the broad interests of the Federation. These publications are intended to provide background information through a review of technical practices and detailed procedures that research and experience have shown to be functional and practical.

Contents

Chapter 3 Guidance Practices

List of Figures

List of Tables

Preface

This publication will provide owners and planners of wastewater collection and treatment systems with strategic planning and decision-making guidance for improving wet weather system performance in the context of National Pollutant Discharge Elimination Systems permit compliance. In addition, this publication is intended to provide federal, state, and local regulators with a better understanding of the best practices for managing wet weather flows and developing implementable plans. Finally, this publication is intended for other interested parties, specifically environmental nongovernmental organizations and those that are generally educated and involved with wet weather issues but may not be aware of current best practices. This update of *Guide to Managing Peak Wet Weather Flows in Municipal Wastewater Collection and Treatment Systems* (2006) is also intended to serve as a fundamental framework for any updated U.S. Environmental Protection Agency sanitary sewer overflows/peak flows approach and will reflect the best current thinking, updated technologies, and improved/proven management approaches on these issues.

This publication was produced under the direction of Daniel W. Ott, P.E., *Co-Chair;* Reggie Rowe, P.E., *Co-Chair;* and Nancy J. Wheatley, *Co-Chair.*

Authors' and reviewers' efforts were supported by the following organizations:

AECOM, Los Angeles, California
Black & Veatch Corporation, Kansas City, Missouri
Brandon Koltz Water & Environmental Consulting LLC, Milwaukee, Wisconsin
Brown and Caldwell, Beltsville, Maryland
CDM Smith, Cambridge, Massachusetts
CH2M HILL, Denver, Colorado
City of Bangor, Maine
City of Brookfield, Wisconsin
City of Moberly, Missouri
City of Reading, Pennsylvania
DC Water and Sewer Authority, Washington, D.C.
East Bay Municipal Utility District, Oakland, California
GBA, Lenexa, Kansas
Hatch Mott MacDonald, Cincinnati, Ohio, and Milburn, New Jersey
Hazen & Sawyer, P.E., Richmond, Virginia
HDR Engineering, Inc., Cleveland, Ohio, and New York, New York

Hydro International, Portland, Maine
Johnson County Wastewater (JCW), Mission, Kansas
KCI Technologies, Inc., Sparks, Maryland
Malcolm Pirnie, Inc., Arlington, Virginia, and Columbus, Ohio
Murray, Smith & Associates, Inc., Portland, Oregon
Optimatics LLC, Chicago, Illinois
Orange County Sanitation District, Fountain Valley, California
Tetra Tech, Ann Arbor, Michigan
URS Corporation, Cleveland, Ohio
Woolpert, Chesapeake, Virginia
Yale University, New Haven, Connecticut

Chapter 1

Introduction

1.0 BACKGROUND

The Clean Water Act's (CWA) goal is to protect and improve the water quality of our nation's water resources. To accomplish this goal, the CWA includes technology-based and water quality-based requirements for discharges to all waters of the United States. Over the 40 years of the CWA, there have been considerable improvements in the health of the nation's water resources, many of which have come from the use of higher levels of technology to treat wastewater and more attention dedicated to optimizing operations and maintenance and upgrading of collection facilities. In addition, regulatory strategies, such as implementation of long-term combined sewer overflow (CSO) control programs to minimize the effects of CSOs, development of *total maximum daily loads* (*TMDLs*) to establish responsibility for pollutant removal, and management strategies, such as implementation of asset management and environmental management systems, have contributed to improvement of overall performance of wastewater collection and treatment systems.

Management of wet weather flows in the wastewater collection and treatment system has remained one of the most intractable problems for wastewater managers, who have struggled to get approvals for affordable wet weather flow management plans. Wastewater planning has traditionally used approaches that addressed primarily dry weather, steady-state conditions. Although facilities planning often considered flow variability and design guidance such as peaking factors, wet weather considerations were not the focus of design, and certainly not as important as optimizing dry weather operations. Moreover, regulatory assessment of the outcome of planning exercises has been inconsistent among agencies and variable over time. For example, some engineering practices that are commonly used as a basis for managing wet weather flows have been acknowledged and included in National Pollutant Discharge Elimination System (NPDES) permits by some permitting agencies, accepted but not included in permit conditions by other agencies, and rejected as a violation of existing regulations at still others. Treatment of similar practices has changed over time as a result of differing interpretations of regulations and pressure from environmental advocacy organizations. With the U.S. Environmental Protection Agency's (U.S. EPA's) introduction of the "Integrated Municipal Stormwater and Wastewater Planning Approach Framework" (2012a), there is the potential that many more wet weather management plans will incorporate stormwater management, in addition to the regulated components of wastewater conveyance and treatment systems. This change will also lead to planning efforts that address facilities owned by public entities other than publicly owned treatment works (POTWs) (POTW is used in this guide to refer to the municipality or agency that holds the NPDES permit for the municipal or regional water resource recovery facility [WRRF]), for example, the municipal public works department with responsibility for stormwater management. Thus, planning in the 21st century involving approvals of wet weather management plans will require proactive management strategies and innovative approaches to achieve success in planning.

Since 2006, the industry has focused on improving the body of professional literature to support POTW planning through joint documents supported by many professional organizations. Examples include *Effective Utility Management; A Primer for Water and Wastewater Utilities, Implementing Asset Management: A Practical Guide* (U.S. EPA et al., 2008), and *Core Attributes of Effectively Managed Wastewater Collection Systems* (APWA et al., 2010).

In addition, U.S. EPA has continued to hold regular dialogues on management of wet weather flow and has worked to implement tools associated with the NPDES process, such as TMDLs and NPDES stormwater permits for municipalities. Despite the addition of these tools, the CSO Control Policy, which has been codified and is now almost 20 years old, remains the last regulatory document that provided a comprehensive methodology to evaluate alternatives for managing peak wet weather flows. U.S. EPA's most recent dialogues have led to the release in May 2012 of U.S. EPA's Integrated Municipal Stormwater and Wastewater Planning Approach Framework (IPF). U.S. EPA's stated intent for the release of the IPF was, "to assist municipalities on their critical paths to achieving human health and water quality objectives of the CWA".

The IPF provides a perspective from U.S. EPA that wet weather flow management should integrate the unique aspects of a community's values and characteristics to develop a plan that will provide the best overall approach to wet weather management. Thus, IPF embraces the concept, discussed in this guide, that wet weather flow management requires the energies and expertise of not only wastewater professionals, but also the broader community, including the additional public entities that manage stormwater systems. All of this supports the proposition that to effectively manage flows and protect human health and the environment, the approach should be holistic and watershed-based. This guide is therefore intended for not only owners, planners, designers, and operators of municipal wastewater collection and treatment systems, but for regulators and other stakeholders involved in managing wet weather flows.

This guide provides a method, or protocol, for POTWs, and partners in an IPF, to be proactive and collaborative in moving forward in planning for wet weather flows. The protocol describes evaluations of wet weather conditions and management alternatives based on risk assessment and performance objectives developed by POTWs and municipal partners. Getting regulatory approval for wet weather management plans has proved challenging over the last few decades, particularly in times of limited and shrinking municipal budgets. The process presented in this guide and the resulting documentation can be used in a dialogue with regulators and other stakeholders to build support for real-world solutions that make effective use of resources in improving water quality.

2.0 PURPOSE AND APPROACH

This guide is primarily intended for POTWs and municipal partners, providing information needed to make decisions on how to improve wastewater system performance during wet weather. It provides a clear and common understanding of what constitutes generally accepted planning and engineering practices and a process for selecting the optimum wet weather management option. Although its focus is primarily technical, the information is provided in the context of regulatory requirements and public policy principles that influence public works decisions.

This guide outlines a systematic process for the analysis of collection and treatment of wastewater flows during wet weather conditions, leading to development of sound and effective practices for municipal facility planning, design, and operation for optimum management of wet weather flows. It is intended to provide guidance on how to consider and approach the many challenging components of wet weather flow management, using examples of effective practices to emphasize and articulate key principles. Readers should be able to use this information to build support for practical solutions that cost-effectively improve water quality.

This guide is an update to the 2006 *Guide to Managing Peak Wet Weather Flows in Municipal Wastewater Collection and Treatment Systems.* The 2006 publication was the result of a project managed by WEF under a Water Quality Cooperative Agreement with U.S. EPA. A project team developed the guide with the advice of a steering committee of WEF volunteer experts and the oversight of a project manager from the U.S. EPA's Office of Wastewater Management. The team also received guidance from a broad-based team of technical reviewers and stakeholders using WEF's established procedures for developing technically sound, balanced materials.

The Federation set the following goals for this project:

- Identify the state of knowledge of wastewater collection and treatment system planning, design, construction, and operation related to wet weather events;
- Synthesize the regulatory requirements regarding treatment approaches and alternatives during wet weather that municipalities should know;
- Establish consensus on this information from experienced system managers, designers, operators, and other recognized experts;
- Solicit input from regulators and other stakeholders on approaches to meet water quality standards and permit requirements; and
- Consider cost-effectiveness and affordability.

The 2006 project team included a steering committee and technical review committee. There were formal face-to-face technical review meetings to review and evaluate the guide content, as well as reviews with industry stakeholders. The goals for the 2006 publication and the strong foundation for the content support the continuing value of the principles and practices that make up this guide.

This 2013 update to the 2006 publication will bring in the best current thinking, updated technologies, and improved/proven management approaches on these issues, as water quality professionals begin to use U.S. EPA's IPF and other excellent references to green and sustainable infrastructure, such as *Planning for Sustainability: A Handbook for Water and Wastewater Utilities* (U.S. EPA, 2012b).

3.0 WET WEATHER MANAGEMENT PRINCIPLES

A POTW's wet weather flow management should be grounded in clear overarching principles. Water quality professionals have learned much about planning design and operation for wet weather facilities since the Clean Water Act of 1972. All must recognize the experience-based criteria that should be applied to planning and designing for wet weather flows. These experience-based criteria may or may not relate to regulatory language and design guidance from the construction grants era. Engineers and other professionals are consequently challenged to be proactive in applying, evaluating, and promulgating practices that reflect decades of experience dealing with actual rather than predicted wet weather flows at WRRFs.

Integration of wet weather planning and design with ongoing system priorities for upgrading and rehabilitation is essential. Whenever possible, wet weather planning should be done holistically as a part of watershed planning when stormwater, habitat, receiving water conditions, floodplain, and other issues are considered and prioritized in context.

The following principles articulate the differences between planning, designing, operating, and maintaining conveyance and WRRFs for typical versus wet weather flow conditions and are the basis for the more detailed guidance that follows:

Principle 1—planning and design evaluations and recommendations must provide a high-quality level of service, protect public health and the environment, and comply with the CWA and other applicable policies, regulations, and permit conditions using best engineering practice.

Principle 2—whenever possible, wet weather planning and solutions should be an integral part of a watershed-based approach to protecting receiving waters, public health, and achieving the CWA's goals.

Principle 3—when planning and designing wet weather flow conveyance and WRRFs, it must be recognized that weather is an uncontrollable but statistically predictable variable. Solutions do not exist that will meet all weather conditions. As climate change continues, statistical analyses will be more challenging, but also more necessary than ever. The goal must remain for collection systems, WRRFs, and stormwater infrastructure to be optimized to improve wet weather performance and reduce risks to public health and the environment, using the best and most current tools available.

Principle 4—actual wet weather flow conditions, collection system, and WRRF performance vary greatly and must be measured or modeled, monitored, and evaluated on an ongoing basis, under the full range of weather conditions.

Principle 5—future flow projections must account for aging infrastructure, increases in infiltration and inflow over time, and actual flow conditions in service areas to be included within the planning horizon, especially if older sewered areas will be connected to new or expanded regional facilities, and considering significant effects on base flow, such as those from economic conditions or from water conservation and reclamation.

Principle 6—the extreme variations in rainfall patterns, topography, groundwater conditions, antecedent moisture conditions, snowmelt/snowbelt issues, and flooding of facilities should be recognized, particularly in light of the effects of climate change. It is more important than ever to evaluate long-term weather patterns and rainfall history and to carefully define extreme conditions—even those that exceed any reasonable capability of municipal facilities to handle. Once identified, such extremes must be documented and included in the planning and design approval process and evaluation of alternatives.

Principle 7—as with all engineering projects and programs, the cost of alternatives should be evaluated, including a realistic assessment of performance standards, costs (including life-cycle costs), benefits, and project optimization so that nonviable alternatives are eliminated from consideration.

Principle 8—non-cost factors can often override technical or financial criteria and must be identified and evaluated as rigorously as the technical performance of alternative solutions, recognizing that most wet weather flow projects must be retrofitted into an already existing, developed community.

Principle 9—wet weather flow solutions are typically long term, capital intensive, and incremental, and should include well-defined, measurable outcomes that can meet regulatory and stakeholder expectations.

Principle 10—wet weather facilities for conveyance and treatment must be effective; performance must be measurable and consistent with monitoring and compliance expectations typically found in NPDES permits. The variability in wet weather conditions and the potential for extreme events must be considered in assigning performance requirements, as well as the infeasibility of measuring and monitoring under all conditions, especially when systemwide flooding is occurring.

Principle 11—proactive operation and maintenance of collection systems and WRRFs must include emergency response to immediately mitigate public health risks, followed by evaluation of environmental effects.

Principle 12—learning by doing is part of the process of managing wet weather flows, and supports adaptive management. Professionals are still learning and developing collection and treatment options for wet weather flows and must encourage continuous learning and information sharing, and the continuous improvement it supports.

4.0 REFERENCES

American Public Works Association; American Society of Civil Engineers; National Association of Clean Water Agencies; Water Environment Federation (2010) *Core Attributes of Effectively Managed Wastewater Collection Systems;* National Association of Clean Water Agencies: Washington, D.C. http://www.wef.org/AWK/pages_cs.aspx?id=1063 (accessed May 2013).

U.S. Environmental Protection Agency (2012a) Integrated Municipal Stormwater and Wastewater Planning Approach Framework; Office of Water and Office of Enforcement and Compliance Assurance; U.S. Environmental Protection Agency: Washington, D.C.; May. http://www.epa.gov/npdes/pubs/integrated_planning_framework.pdf (accessed May 2013).

U.S. Environmental Protection Agency (2012b) *Planning for Sustainability: A Handbook for Water and Wastewater Utilities;* EPA-832-R-12-001; Office of Water; U.S. Environmental Protection Agency: Washington, D.C.; Sept. http://water.epa.gov/infrastructure/sustain/upload/EPA-s-Planning-for-Sustainability-Handbook.pdf (accessed May 2013).

U.S. Environmental Protection Agency; American Water Works Association; Association of Metropolitan Water Agencies, National Association of Clean Water Agencies; National Association of Water Companies; Water Environment Federation (2008) *Effective Utility Management; A Primer for Water and Wastewater Utilities, Implementing Asset Management: A Practical Guide.* http://nepis.epa.gov/Exe/ZyNET.exe/P10053BJ.TXT?ZyActionD=ZyDocument&Client=EPA&Index=2006+Thru+2010&Query=&SearchMethod=1&QField=&XmlQuery=&File=D%3A%5Czyfiles%5CIndex%20Data%5C06thru10%5CTxt%5C00000010%5CP10053BJ.txt&User=ANONYMOUS&Password=anonymous&Display=p%7Cf&DefSeekPage=x&MaximumPages=1&ZyEntry=1&SeekPage=x&ZyPURL# (accessed May 2013).

Chapter 2

Wet Weather Management and Planning Approach

1.0 WET WEATHER MANAGEMENT FRAMEWORK

1.1 Introduction

Managing wet weather flows to protect public health and the environment and ensure Clean Water Act (CWA) compliance requires publicly owned treatment works (POTWs) to integrate the planning, design, and operations and maintenance (O&M) of wastewater collection and treatment systems. As regulation of municipal stormwater systems has been introduced in the last several years, communities may now look to adding management of municipal stormwater infrastructure to wet weather flow management. The potential for this integrated planning has been enhanced by the U.S. Environmental Protection Agency's (U.S. EPA's) release of the "Integrated Municipal Stormwater and Wastewater Planning Approach Framework" (IPF) (2012). The goal is to improve performance of all wastewater and stormwater infrastructure by preparing for anticipated wet weather conditions that affect the O&M of the entire municipal wastewater, and perhaps, stormwater systems, while recognizing that severe, unmanageable events and conditions are likely to occur.

This guide's approach is to use assessments of technology and technical information to identify alternatives for improving performance and practical limitations of designing and operating facilities during extreme events. It also uses values-based risk management to incorporate community values and expectations for level of service. Finally, it is grounded in wet weather management principles, provided in Chapter 1, that challenge water quality professionals to be proactive in applying, evaluating, and promulgating practices that reflect the decades of experience in wet weather flow management. By working through this process, a POTW can develop capital programs and management systems for ongoing operations based on good engineering practices and cost/benefit considerations that enhance the likelihood that wet weather noncompliance events are limited to those that are unavoidable. Although the process itself is not a compliance assessment, it can support discussions with regulators about the POTW's performance and available alternatives, potentially including management of municipal stormwater infrastructure that might further minimize risk to public health and the environment. To have the best chance of winning regulatory acceptance for a plan that meets the community's values and needs, it is critical that the documentation highlight the comprehensive nature of the evaluation and demonstrate broad community support for the expected planning process and preferred alternative.

This guide describes successful wet weather management practices and processes; provides methods to characterize the performance of wet weather management alternatives; provides a framework for holistically identifying and assessing site-specific risks in terms of service delivery, public health, environmental, or other factors; and suggests an approach for planning, selecting alternatives, and implementing wet weather management strategies.

1.2 Wet Weather Regulatory Framework

Before beginning any planning process for wet weather flow management, it is important to understand the regulatory context for that planning. The POTW and its planning partner must have a clear understanding of the regulatory framework in which POTWs function. The CWA and its attendant regulations are the basis for that framework. The National Pollutant Discharge Elimination System (NPDES) program sets forth requirements to minimize or eliminate discharge of pollutants by POTWs to protect and enhance the quality of our nation's waters. Wet weather management, however, has resulted in particular challenges for municipal wastewater system planning, design, and management, including determining compliance with CWA requirements. Regulatory approval for wet weather control plans, combined sewer overflows (CSOs), and other wet weather issues, have been far from easy or quick. The question of what "compliance" and "in all conditions" mean has been a topic of discussion for decades—in U.S. EPA federal advisory committees; as part of negotiations and implementation of consent decrees; as stormwater regulations have been developed and adopted; and as innovation and technology have added to potential wet weather controls. However, as evidenced by U.S. EPA's failure to adopt a new sewer system overflow (SSO) rule, a clear answer to the question of what "compliance" and "in all conditions" means still does not exist. The approach outlined in the next chapter is intended to provide the POTW and its planning partners the best opportunity to demonstrate that a plan does meet that uncertain standard.

The following is a regulatory framework for assessing compliance with CWA requirements, based on U.S. EPA's interpretation of the CWA and regulations, as well as court rulings:

- Clean Water Act provisions and NPDES requirements associated with standards and permitting:
  - Minimum technology standards apply to all discharges regulated by the CWA [FWPCA §301(b)(1)(B)]:
    - For POTWs, effluent discharges must meet "secondary treatment" performance criteria, and
    - The CWA definition for a POTW includes the collection system used to convey wastewater to the permitted water resource recovery facility (WRRF);
  - Additional treatment or controls beyond the minimum technology standard (e.g., secondary treatment for POTWs) apply when effluent discharges could "cause or contribute to" impairment of a waterbody [FWPCA §302(a)];
  - State or tribe adoption of water quality criteria to protect beneficial uses;
  - Facility plans that assess service area requirements and other factors to ensure that facilities are properly sized for an appropriate range of flow conditions;
  - A POTW submission of an NPDES permit application that includes design and operating conditions and parameters, such as anticipated bypasses for the wastewater system;

- Issuance of an NPDES permit that must include effluent limits and other terms to ensure that discharges will not cause or contribute to impairment of water quality and prevent designated uses from being attained; and
- Additional permit requirements, such as O&M requirements and bypass and upset provisions of CWA regulations.

- U.S. EPA has used the bypass regulation to object to WRRFs routing flows around secondary treatment units to avoid damage to those treatment units during peak flow conditions. Application of the bypass regulation sets a "no feasible alternatives" standard to allow exceptions to the prohibition of bypasses. *Proposed EPA Policy on Permit Requirements for Peak Wet Weather Discharges from Wastewater Treatment Plants Serving Sanitary Sewer Collection Systems* (U.S. EPA, 2005) was never finalized, nor has any other guidance or regulation been forthcoming. On March 25, 2013, the 8th Circuit Court of Appeals ruled in Iowa League of Cities v. U.S. EPA [No. 11-3412] that U.S. EPA had no statutory authority to prohibit blending, along with several other procedural rulings. On May 9, 2013, U.S. EPA petitioned for rehearing of the case in by the 8th Circuit. If the decision is not overturned in the rehearing, it would be applicable to POTWs under the jurisdiction of the 8th Circuit and would influence compliance evaluation of blending practices.
- The Combined Sewer Overflow Control Policy, which has been codified, addresses wet weather flow requirements in combined sewer systems and requires nine minimum control measures for managing existing systems, with specific long-term planning and design approaches to control wet weather flows; and
- National Pollutant Discharge Elimination Systems compliance determinations that include those based on whether a specific discharge is permitted and whether it meets technology and water quality-based standards as well as other CWA requirements and the associated regulations.

In the last decade, permitting authorities and enforcement staff have begun to identify management programs such as collection system capacity management, operations, and maintenance (CMOM) in NPDES permits and consent agreements. Although the programs have not been refined, developed, and approved as policy or guidance and are not coordinated or linked with the technology and water quality-based standards of the CWA, they are being widely used. In addition, the guidance documents developed by the professionals communities, such as the *Core Attributes of Effectively Managed Wastewater Collection Systems* (APWA et al., 2010), have served as the industry's contribution to best practices for improving system performance.

In May 2012, U.S. EPA released the IPF (U.S. EPA, 2012) to "assist municipalities on their critical paths to achieving the human health and water quality objectives of the Clean Water Act", while still requiring that NPDES permittees fully comply with all CWA requirements. In the IPF, U.S. EPA recognizes that integrated planning for wastewater and stormwater management can lead to more cost-effective capital programs. Moreover, it embraces use of green, sustainable, and innovative technologies. Finally, U.S. EPA believes that the IPF can be implemented within the existing NPDES regulatory structure. Potential benefits to NPDES permittees in using this approach include the following:

- Establishes the ability to prioritize the order of capital projects based on benefits;
- Allows for adaptive management to review, evaluate, and change future projects based on performance of earlier projects; and
- Encourages the use of green and innovative technologies, even when performance data may not be readily available to support strict compliance.

The IPF offers POTWs the potential for more flexibility in identifying and establishing the priority for wet weather flow management infrastructure alternatives. Nevertheless, POTWs must continue to manage the performance of treatment and collection systems during wet weather to provide uninterrupted service, protect their facilities, and comply with their NPDES permit conditions, thereby protecting public health and the environment.

1.3 Wet Weather Flow Management Planning

With this framework in mind, the next topic is the basics of wet weather flow management planning. The traditional approach to wastewater planning primarily focused on steady-state operations and meeting water quality standards, which were based on dry weather conditions with planning for some flow variability, such as increased wet weather flows. Experience has demonstrated that wet weather flow management must address site-specific conditions, including highly variable flows and discharges to waterbodies that are affected by wet weather sources and conditions other than those that are WRRF-related. Specific limitations in the traditional approach include the following:

- It may not sufficiently identify and provide treatment approaches that handle wet weather flows most effectively. This may create a bias for adding storage facilities or process enhancements without adequately considering overall process limitations, such as impaired performance of WRRFs under stressed conditions during high flow (wet weather) events. It may also lead to oversizing process facilities with other adverse operational consequences.
- It does not recognize that overflows from the collection system may occur even if the system is well designed and maintained because of extreme events outside the planning, design, and operational performance

parameters of the system and beyond well-established engineering practices and standards.

While the regulatory framework presented in the previous section theoretically provides a baseline for planning criteria, there is a disconnect between strict interpretation of regulatory compliance and the need for clear planning criteria to develop workable and cost-effective wet weather management plans. Perhaps the best example of this is the often asked question, "How many SSOs are allowable?" The regulator's answer is invariably "None". This disconnect leaves POTWs in a regulatory quagmire that makes planning difficult.

Generally accepted engineering and economic practices identify the best technology, along with specific design options and management practices. These practices are a foundation for planning and form the basis for the processes in the following sections. Together, these processes outline a strategy to identify wet weather management alternatives and develop objective support for evaluating and selecting alternatives. A process to establish appropriate limits of system performance is combined with a values-based risk management approach that uses community input to establish the underlying values.

In sum, this a strategic approach to putting the justification for wet weather management plans in the context of the community and provides the community's answer to the question of what it expects to "how good is good enough" for its waterways during wet weather events.

1.4 Wet Weather System Performance

Next in the process, POTWs must understand the underlying nature of performance exceptions. They must consider the difference between performance exceptions caused by system failures and those caused by system deficiencies. They are fundamentally different.

- A *system failure* occurs when a system no longer performs its intended function because of extreme weather or other extraordinary events beyond design parameters (e.g., Hurricane Sandy), and additional controls or treatment are not feasible or economical based on generally accepted engineering practices; and
- A *system deficiency* exists when a system's performance is impeded from its intended function but capital or management systems improvements to the collection and/or treatment system may be designed and built affordably, consistent with generally accepted engineering practices, which would eliminate or reduce negative effects (e.g., a system deficiency can be corrected by applying CMOM best practices).

An illustration of the decision logic to determine responses to system deficiency and failure is shown in Figure 2.1. The logic flowchart shows the importance

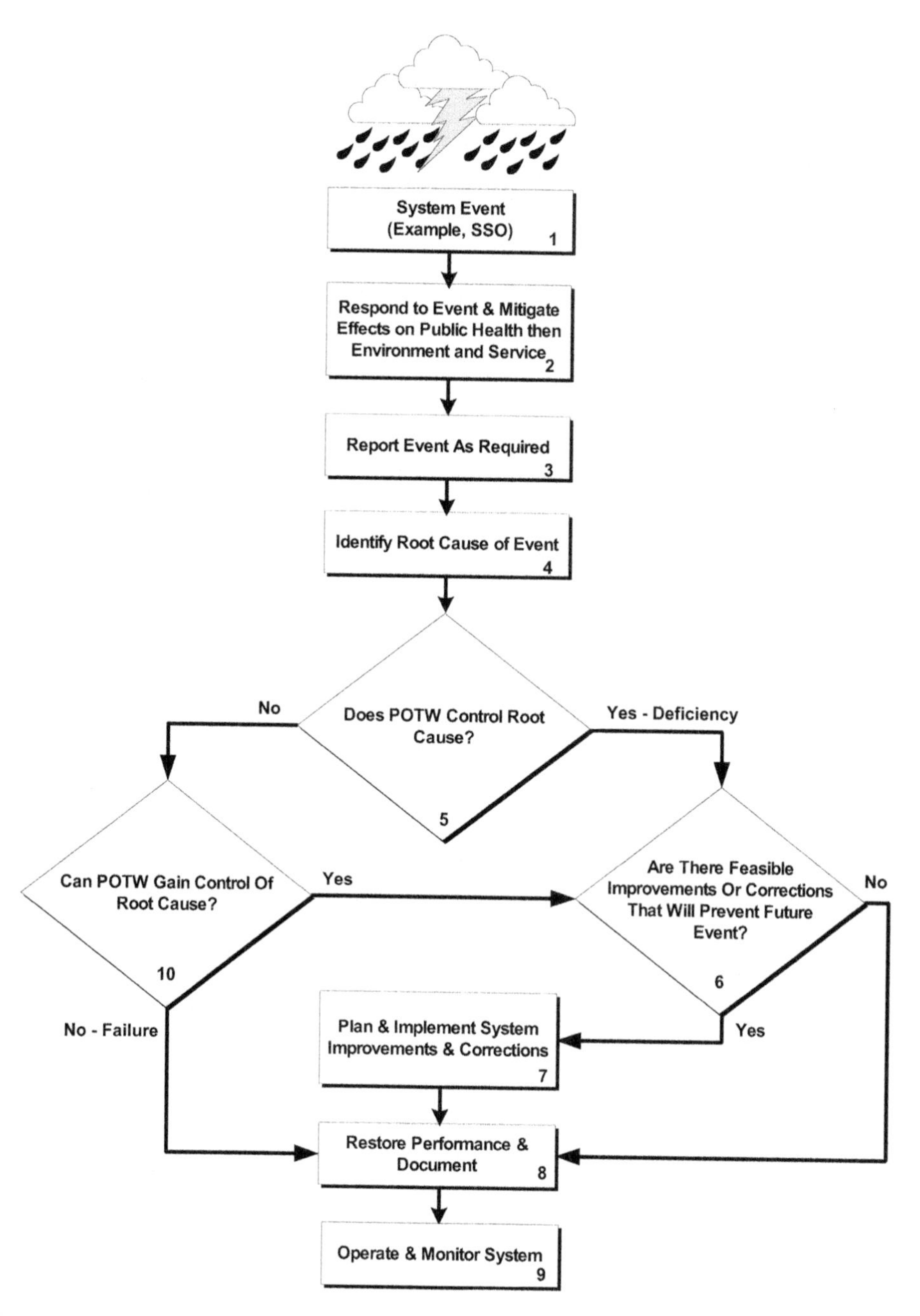

FIGURE 2.1 An example of decision logic to determine system deficiency and failure.

of knowing the root cause of the performance exception and whether the POTW can control the root cause, as shown in Box 4, Diamond 5, and Diamond 10, respectively. A system failure exists is when the response to Diamond 6 is "No" and potentially when the response to Diamond 10 is "No".

System deficiencies may be associated with either inadequate facilities planning, poor O&M practices, or older systems that were properly designed and built to past standards that do not meet current expectations. Wet weather incidents (bypasses at the WRRF or SSOs in the collection system) would not have occurred if current generally accepted design and O&M practices were in place and followed. System failures represent the practical limitations of wastewater treatment and collection systems to operate in extreme wet weather or other conditions that exceed generally accepted engineering design and construction practices. Table 2.1 lists several common performance exceptions, their root causes, and whether they were system deficiencies or system failures. These examples show that system deficiencies are correctible and may or may not be avoidable. System failures may or may not be correctible and are unavoidable. An unavoidable event or a system failure is not necessarily an authorized performance exception. For example, an SSO that reaches the waters of the United States is prohibited unless it complies with the requirements of the CWA.

Professional engineers and others involved with management of wet weather flows recognize that gray areas exist between system deficiencies and failures that relate to the innovation and evolving success or failure of new technologies and approaches to controlling wet weather flows. A technology's performance may vary from one location and set of conditions to another, and evaluating and monitoring system performance in episodic intervals associated with wet weather events can be challenging.

The approach recommended in this guide is a values-based risk analysis, combined with generally accepted engineering and economic practices, to establish wastewater collection and treatment planning for wet weather flow conditions. The steps to apply a system performance analysis during planning include understanding the CWA and regulatory requirements; establishing the POTW's mission and goals; developing levels of service performance objectives for wet weather conditions; and assessing system failures and system deficiencies. Generally accepted engineering and economic practices will identify the best technology, along with specific design options and management practices. The performance-based risk analysis provides information on the effects of various alternatives, allowing a POTW to determine if there are *feasible alternatives*, that is, if additional investment and/or improved O&M can reduce risks to public health and the environment during wet weather operations of wastewater systems.

This guide enables POTWs, regulators, and other stakeholders to use similar language (i.e., terms) to evaluate whether performance exceptions were avoidable and if further facilities and controls are required to manage wet weather flows. The approach it recommends can improve customer service, garner stakeholder support, provide appropriate protection from failures, and minimize environmental and public health risks. It also can improve a POTW's ability to comply

Table 2.1 Causes of collection system performance exceptions.

Cause of performance exception	Root cause of performance exception	Considered avoidable?	Considered unavoidable?	Considered a deficiency?	Considered a failure?
Blockage	Fats, oil, and grease built up over time	Y	N	Y	N
	Debris entered system suddenly or in short time span	N	Y	Y	N
	Object from third-party construction activity penetrated sewer system	N	Y	Y	N
System collapse	Utility owner's heavy equipment crushed sewer	Y	N	Y	N
	Unpermitted earthen overburden was placed in sewer easement	N	Y	Y	N
	Pipe corroded	Y	N	Y	N
Flow exceeded capacity	Extraneous infiltration/inflow during a rare wet weather event (far above-average long-term return frequency, intensity, and duration); widespread flooding	N	Y	N	Y
	Extraneous infiltration/inflow during an unusual wet weather event (above-average long-term return frequency, intensity, and duration); management alternatives were not evaluated, nor were feasible improvements implemented	Y	N	Y	N
	Extraneous infiltration/inflow during an unusual wet weather event (above-average long-term return frequency, intensity, and duration); management alternatives were evaluated and feasible improvements were implemented	N	Y	N	Y
	Flow in satellite system with poor operations and mainenance practices exceeded regional trunk system capacity	Y	N	Y	N
	Private-property laterals let extraneous water into the system	Y	N	Y	N

(*continued*)

TABLE 2.1 Causes of collection system performance exceptions (*Continued*).

Cause of performance exception	Root cause of performance exception	Considered avoidable?	Considered unavoidable?	Considered a deficiency?	Considered a failure?
	Treatment facility operator limited flow through biological treatment units to avoid losing biomass during a rare wet weather event (extremely above-average long-term return frequency, intensity, and duration)	N	Y	N	Y
Power failure at pumping station	Electrical storm interrupted power for 24 hours at a significant station that lacks backup power generators or electrical feed	Y	N	Y	N
	Lightning burned up two of the station's three large pump motors	N	Y	N	Y
	Tornado downed power transmission feeds and blocked station access for 2 days	N	Y	N	Y

and demonstrate compliance with regulatory requirements, while recognizing the limitations of physical systems and cost/benefit considerations. It enables POTWs to focus on the goal of continually improving performance, consistent with the overall goal of the CWA and community goals to protect and enhance our water resources.

2.0 PERFORMANCE OBJECTIVES USING VALUES-BASED RISK MANAGEMENT APPROACH

A successful POTW is one that establishes a set of values and attains its mission. It is important that the POTW values are properly perceived in its corresponding level of service (LOS) expectations of the community, are captured in its mission statement, and are developed into strategies and programs to accomplish them. Equally important in the implementation process are performance measurement tools to measure performance and make necessary adjustments to stay within performance tolerances and LOS objectives.

For purposes of presenting the basics of this approach, the text will focus on the leader of the process, which will be referred to as the POTW. Those POTWs that are regional agencies will typically have satellite collection systems owned and operated by municipalities. In this case, and also if the POTW plans to use the IPF to incorporate stormwater management in the planning process, multiple municipal and other public partners will be involved in the planning. The implementation of the process discussed below will be driven by the respective roles and responsibilities of each party specific to the region or watershed, and participation by other municipal entities should be agreed to before beginning the process.

The relationship among the POTW's mission, values, goals, LOS performance objectives, and performance measures as used in this guide is shown in Figure 2.2. Figure 2.2 illustrates that individual goals may have more than one LOS objective and corresponding performance measure and target. These organizational concepts must be linked to support each other. They are defined as follows:

- The POTW's *mission* is its expression of its purpose in the community and the values it wants to protect.
- The POTW's *values* are simply what are important to the POTW.
- The POTW's *goals* are the stated results that help the POTW attain its mission. Goals are typically expressed as nouns or subjects, such as "satisfied

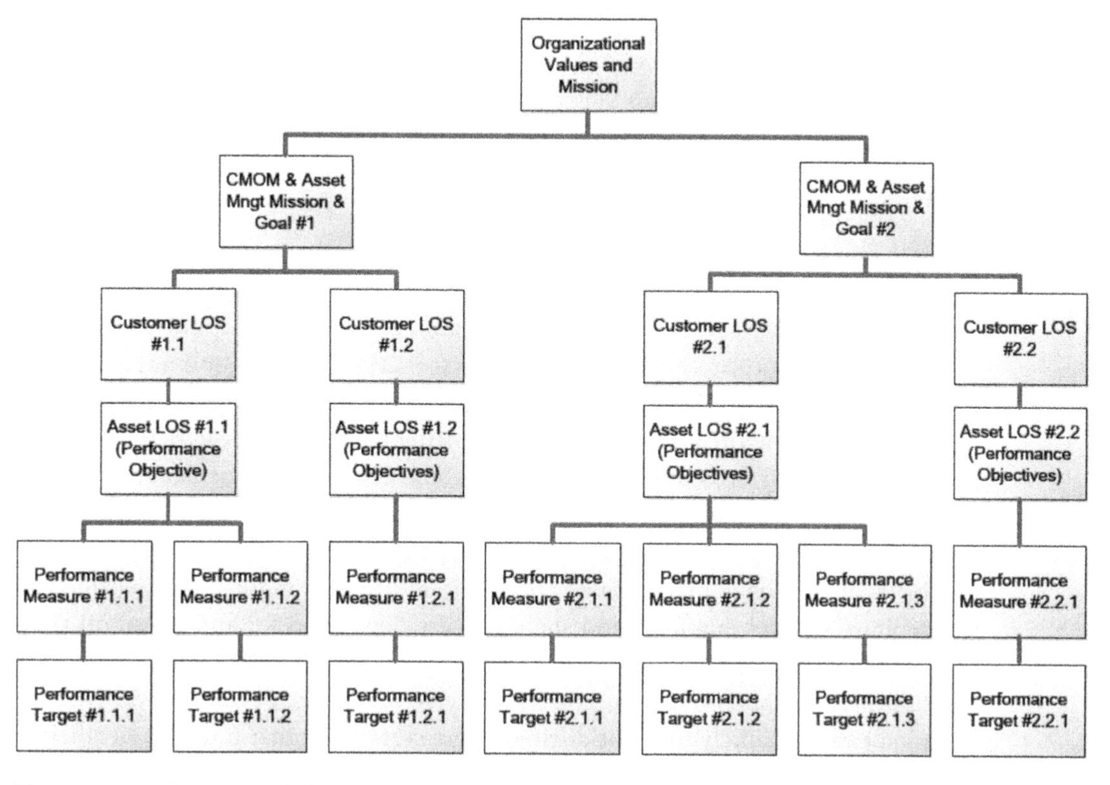

FIGURE 2.2 Framework for communicating business strategy implementation.

customers" or "regulatory compliance". Goals, values, and objectives are often used interchangeably.

- The POTW's *Level of Service* is the type and quality of service provided by the POTW or by the assets as a whole within the wastewater collection and treatment systems. There are two associated types of LOS:
    - *Customer LOS* represents what the customer's expectations are from the POTW in terms of services provided, and
    - *Asset LOS* represents what results are needed from the asset's performance or management policies to meet customer expectations and organizational goals/objectives. Asset LOS is often referred to as performance objectives.
- The POTW's *performance objectives* (also referred to as asset LOS) further break down and explain the components of the POTW's goals and specify the LOS outcome to be achieved. Objectives typically are expressed as an action item and begin with a verb, such as "minimize service disruptions" or "comply with permits and regulations".
- The POTW's *performance measures* (also referred to as performance indicators) explain how each LOS objective will be quantitatively or qualitatively measured and tracked and by what metric, criteria, or parameter; for instance, "*X* number of", "meters of", or "volume per".
- The POTW's *performance target* is the numerical value or saleable narrative description by which performance measures are compared.

Planning for POTW capacity requirements has traditionally been based on separate evaluation and planning for collection and treatment within municipal systems. To successfully address wet weather flows in a system, evaluation and planning efforts should be integrated. Figure 2.3 illustrates an integrated facilities planning process and shows the steps for setting capacity and implementing selected flow control alternatives.

The integration of collection and treatment planning begins with the definition of performance objectives (see Figure 2.3, box "Establish performance objectives for collection and POTW system"). Traditionally, POTW performance objectives have been based on guidance for the design of conveyance and POTWs, which focus on the loads that must be conveyed and treated regularly. Consequently, sanitary engineering texts and standards are developed for and tested on average waste strength and flows, and extrapolated to peak daily flows, maximum hourly flows, and other common conditions—and may not consider the uncommon extremes likely to occur during severe storms or other unusual weather.

Wet weather performance necessarily focuses on extremes (peak days or peak instants) that may be neither defined nor tested in traditional sanitary engineering

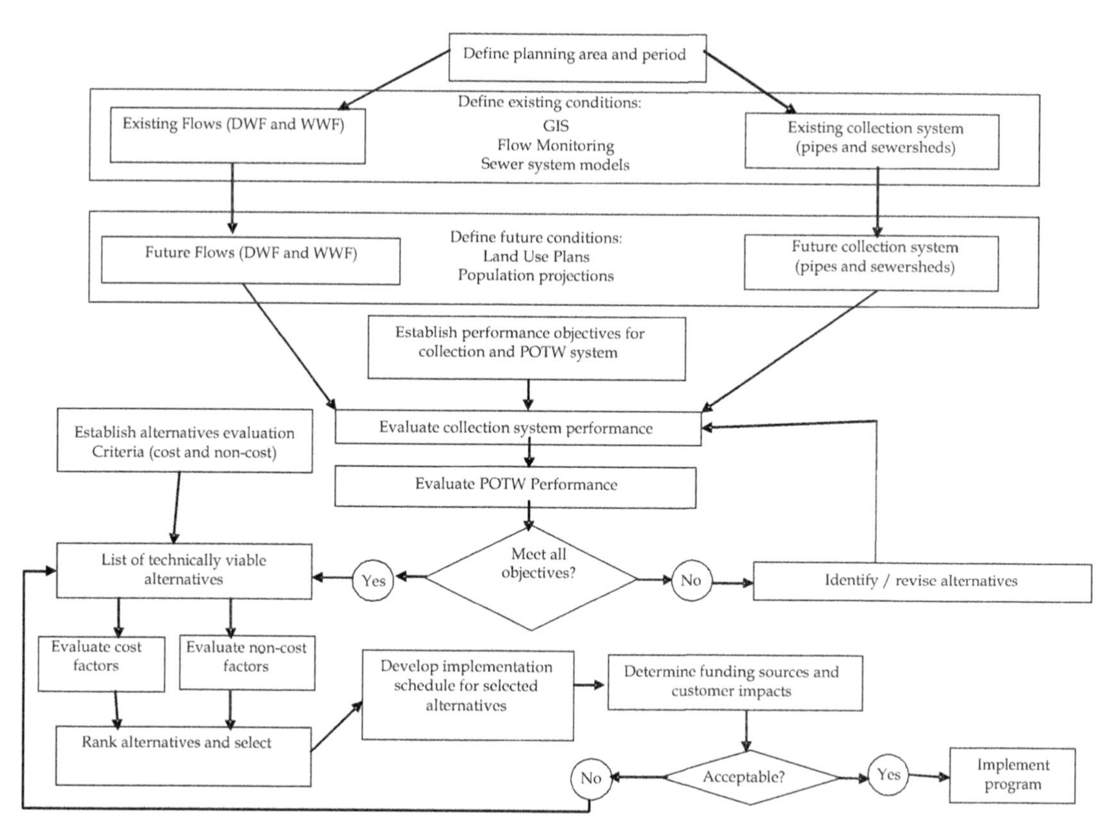

FIGURE 2.3 Facilities planning flow chart (DWF = dry weather flow; WWF = wet weather flow; and GIS = geographic information systems).

guidance. Biological treatment systems, as well as other components of WRRF processes, are not amenable to rapid increases and decreases in flow and loads, making it difficult or impossible to provide effective secondary treatment for extreme flows. As a result, performance objectives are often inadequately developed to properly manage high flows and minimize overflows and performance deficiencies. System capacity has been typically undersized for wet weather events and system discharges have become regular occurrences.

Definition of performance expectations under wet weather conditions requires extrapolation beyond traditional sanitary engineering guidance. Concepts traditionally developed for flood flow management, not wastewater management, provide precedent for addressing sanitary conveyance high flow extremes. Modern flood management establishes broader performance objectives that are related to the predicted probability (frequency) of occurrence and severity of performance failures, an approach that needs to be applied similarly to managing wet weather flows in POTWs. During wet weather, POTWs must respond to the uncertainty of the weather (combined with the antecedent moisture and temperature conditions), requiring design and operation based on probabilistic concepts similar to those governing flood and stormwater controls.

This guide outlines methods for applying proven risk management concepts to the definition of wet weather performance objectives for collection and treatment facilities. The concepts are based on those described in *Capital Planning Strategy Manual* (AwwaRF, 2001) and *Managing Public Infrastructure Assets to Minimize Cost and Maximize Performance* (NACWA, 2002). Figure 2.4 demonstrates an enhancement to the traditional planning process by including six steps that make up the risk management process as presented by this guide. Some literature will include more or fewer steps depending on the intent of the literature's discussion. These six steps are how a POTW establishes and works to achieve its performance objectives. The other activities in Figure 2.4 follow concepts of the traditional facilities planning process and are consistent with successful management programs based on performance assessment and targeted improvements.

The six steps (numbered boxes shown in Figure 2.4) in the values-based risk management approach for collection and treatment systems are further described below.

1. Define and score or rank community and system values,
2. Define and assign wastewater system risks to achieving defined values,
3. Define objective measures of wastewater system-related risks to the defined values,
4. Estimate values achieved at various levels of risk management,
5. Select target level of performance risk, and
6. Adopt performance objectives for each system.

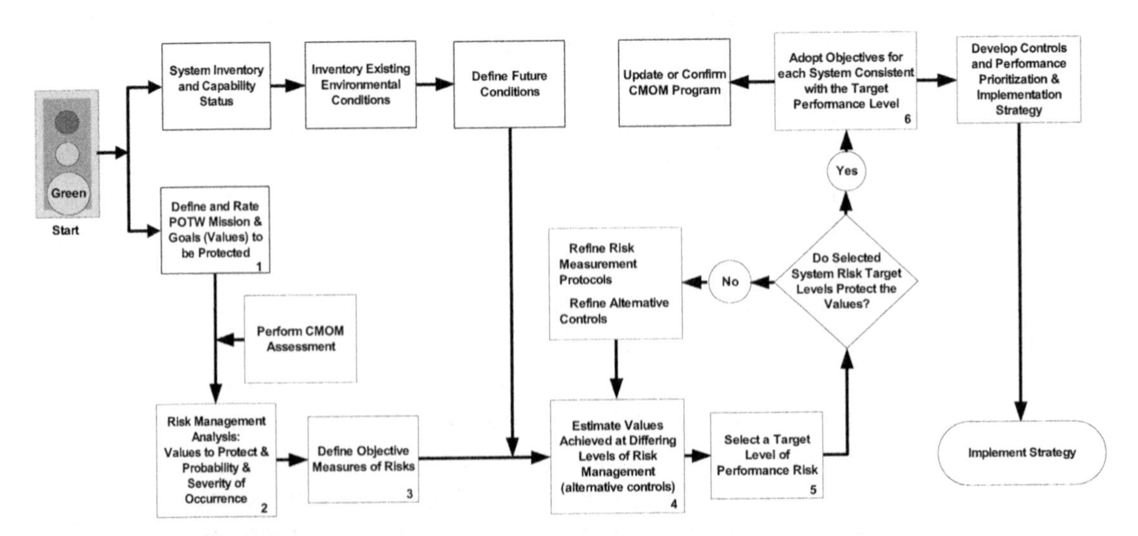

FIGURE 2.4 A process for implementing practices with values-based risk approach.

This is a structured process that combines stakeholder input on the importance of potentially competing community values with the technical and scientific approaches to comparing activities that compete for limited resources. Because this is a community-based process, it is important to put the term "stakeholder" in context. Publicly owned treatment works routinely include public participation in all types of activities, and the participants are typically collectively the "stakeholders". Common members of this group are elected officials; community leaders; interested members of the public; members of advocacy groups; professionals who may or may not be paid for their work; and, for most POTW activities, regulators. Also note that regulators are also decision-makers when it comes to regulatory compliance. An example of where this structured process combining the technical work with community values has been successfully used is the Louisville and Jefferson County Metropolitan Sewer District (Swanson and Kraus, 2009).

The outcome of following the risk-based process is a regulatory compliance approach that achieves performance objectives consistent with stakeholder-defined community values and LOS expectations. This process enables POTWs to define and adopt feasible, locally applicable solutions to meet wet weather and water quality needs.

This approach recognizes that risks cannot be eliminated, but they can be managed. Risk management theory recognizes that any risk is composed of the following two components:

- The likelihood (frequency) of that risk and
- The consequence (or severity) of that risk.

The level of risk is a combination of the risk level identified for each component separately, as illustrated in Figure 2.5. In completing the risk analysis, numeric values would be assigned to each axis. In Figure 2.5, risks that have high likelihood and high consequences (e.g., the health risk from a CSO discharging to a beach during small storms) are in the upper left corner. Risks with low probability and few consequences (e.g., a control gate fails to close the storage tank, causing overflow to a shipping tunnel) are in the lower right corner. The risk to the public health value is real in both cases, but the significance of the risks and the priority of their control differ.

This six-step approach formalizes the consideration of multiple values and risks, allowing an open discussion and traceable documentation of how risks were identified and prioritized. The method for identifying, analyzing, and ultimately selecting processes and facilities to meet wet weather performance objectives is linked to the technical practices found in Chapter 3. In other words, the technical practices are the engineering solutions that a POTW uses to successfully improve its performance during wet weather events.

A description of each of the six steps follows. The discussion below focuses on the POTW and the wastewater collection and treatment system. With the IPF

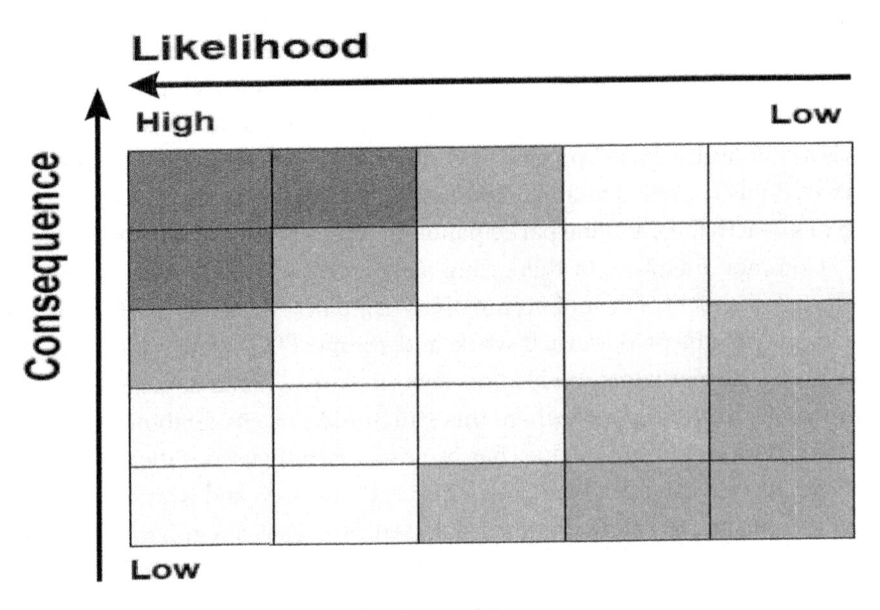

FIGURE 2.5 Risk components and relationship.

making watershed planning more attractive and offering potential for more cost-effective alternatives, planning teams will be more likely to go beyond wastewater service providers and facilities by adding stormwater management to the planning process. The basic approach discussed in each of the six steps is not bounded by the type of agency—wastewater versus stormwater or specific facilities—and can be adapted for any planning group.

2.1 Define and Score or Rate Community and System Values

Step 1 is to identify and rate, either numerically or by prioritization, the values to be protected against results of performance deficiencies or failures. These values are the ultimate benefits or goals to be achieved through improved management and performance. Strategic plans for public wastewater utilities commonly define the applicable values. Examples include CWA compliance; permit or regulatory requirements; public health; environmental well-being; economic growth; consistent service; and protection of POTW assets.

Steps to defining these values include the following:

- Survey stakeholder values and score or rate them on a defined scale,
- Document value descriptions and their scored rankings,
- Confirm these values with selected stakeholders,
- Reach consensus on the values, and
- Document the values.

In most communities, POTWs recognize and protect some variation of the following values (listed alphabetically):

- Affordable POTW rates;
- Community aesthetics;
- Economic growth and diversity;
- Environmental protection;
- Protection of real property;
- Public health;
- Quality service, including protection from backups;
- Regulatory and permit compliance;
- Service expansion, when needed; and
- Uninterrupted service.

These example values often form a direct basis for subsequent LOS or performance objectives.

2.2 Define Wastewater System Risks to Achieving Defined Values

Step 2 is to identify the treatment and collection system risks that could prevent the POTW from achieving the defined values. Numerous factors, directly related and unrelated to a wastewater system, result in challenges to a POTW's achievement of the defined values. In the case of achieving coliform standards, for example, studies have identified risks and the portion of those risks that were wastewater system-related. Examples of these risks include ingestion of pathogens caused by spills from the conveyance or treatment system; ecological stress in the receiving waters; rate stress for customers; connection moratoriums resulting from noncompliance; basement backups; and costs exceeding community ability to pay.

Steps to defining the wastewater system-related performance risks to the values include the following:

- Brainstorm the performance threats to each value (e.g., ingestion of pathogens in recreational waters threaten public health; poor water quality threatens aquatic life; burdensome rate levels; connection moratoriums; basement backups; damaged POTW assets);
- Hypothesize, perhaps using historic data or updated statistical planning data for weather conditions, the potential for wastewater system effects on each value (e.g., untreated sewer overflow that reaches recreational waters would increase risk to public health; untreated wastewater overflows

could cause poor water quality; correcting existing wet weather capacity problems would cause POTW rates to increase; taking no action to correct existing SSOs could lead to a consent decree or other enforcement action that could include a connection moratorium; basements in certain service areas may flood in wet weather; POTW assets could become inoperable if subjected to flows beyond design capacity);

- Evaluate and document the nature of each wastewater system-related performance deficiency or failure (e.g., untreated sewer overflows discharge pathogens and could affect public health resulting from increased contact; non-point sources of pathogens could affect public health where no wastewater is discharging; untreated discharges and other hydrologic conditions [low flows] could depress dissolved oxygen; rates could increase on an annual basis; wastewater backups to homes and businesses could increase in certain areas during wet weather);
- Evaluate and document the likelihood of each wastewater system-related performance deficiency or failure (e.g., the probability of sewer overflow from a combined sewer is high; likelihood of sewer overflow from a sanitary sewer is lower; likelihood of burdensome POTW rates is high; likelihood of connection moratorium is higher or lower based on regulatory history of permit authority; likelihood of basement flooding increases with wet weather intensity and duration);
- Evaluate and document the consequence of each wastewater system-related performance deficiency or failure (e.g., severity from treated wastewater discharge is low; consequence from partially treated wastewater discharge is intermediate; consequence from undiluted wastewater discharge is very high; consequence of wastewater-related illness is intermediate with untreated sanitary discharges; rate increases may be severe to low-income households; risk consequence is high for local businesses that use large amounts of water; consequence of basement flooding is high, but the likelihood of flooding may be low); and
- Select and implement a process to "score" the wastewater system-related risk factors.

Examples of risk factors commonly identified by POTWs are summarized in Table 2.2.

2.3 Define Objective Measures of Wastewater System-Related Risks to the Defined Values

Step 3 is to define objective measures that evaluate system-related risks based on defined community values. Developing scalable, objective measures of each risk enables POTWs to track their progress, thereby encouraging those activities that

TABLE 2.2 Community values and collection system-related risks.

Community value	Common sewerage-related risks
Public health	Beach closures
	Shellfish-bed closures
	Illness
	Sewer backups into buildings
	Sanitary sewer overflows
	Combined sewer overflows
	Catastrophic loss of treatment units
The environment	Nutrient discharges contributing to eutrophication
	Organic discharges depressing dissolved oxygen levels
	Solids discharges restricting access
	Floatables discharges affecting aesthetics and safety
Real property	Wastewater backups causing property damage
	High-velocity wastewater washing out equipment
	Wastewater flooding destroying equipment
	Sewer collapse hindering transportation
	Sewer collapse weakening building foundations
	Street flooding making roads impassable
Community aesthetics	Septic wastewater generating odors in populated areas
	Unscreened wastewater overflows depositing objectionable materials (floatables) in visible areas
Economic prosperity (meeting expansion needs, affordable rates)	Perceived lack of sewer capacity results in building moratorium
	Public health threatened by failing onsite systems
	Cost of capital improvements to treatment works
	Cost of capital improvements to sewer systems
	Personnel costs of better construction inspection
	Personnel costs of better monitoring and documentation
	Higher operation and maintenance costs to implement capacity, management, operations, and maintenance program
Compliance with regulatory requirements	Misunderstanding of applicable regulations and definition of extreme wet weather event
	Changes in regulatory requirements
	Technically infeasible permit conditions
	Uncertainty about performance in all conditions

improve the performance and discouraging ones that increase failures or exacerbate deficiencies. The measures should be scalable, not pass or fail, so POTWs can track progress toward the goal, even if achieving the ultimate goal is infeasible. Examples include time public is exposed to certain pathogen levels; fish and aquatic organism diversity index; percentage of income to support wastewater system corrective programs; number of SSO violations; backups per month per inch of rain; and frequency of maximum capacity utilization.

Steps to defining objective measures of the wastewater system related performance risks include the following:

- Brainstorm measures of each wastewater system related performance risk (some measures may be numeric, whereas others objectively categorize the related likelihood and consequence factors).
- Identify and document protocols for consistently evaluating effects on each wastewater system-related performance risk. Units of measurement must be practical, economical, and easy to maintain. This is where sampling, monitoring, data evaluation, and modeling tools should be considered.
- Pilot test or peer review of each performance measure. It is important to test that the protocol consistently measures the existing performance or condition and the condition or performance under alternative future control measures.
- Seek stakeholder endorsement of the performance and risk measurement protocol.
- Refine and document the protocol for performance and risk measurement.

Although the manager responsible for demonstrating progress is primarily responsible for defining meaningful, useable performance measures, the process requires the cooperation of a wide range of stakeholders because it is essential that the measures objectively reflect progress (or lack of progress) toward reducing the perceived risks to various stakeholder goals. Experience indicates that an open, cooperative approach to identifying performance measures can result in broad endorsement of the metrics themselves and the progress they reflect.

2.4 Estimate the Values Achieved at Differing Levels of Risk Management

Step 4 is to look at several levels of risk management to assess performance based on values. Examples include the effect of reducing overflows to once per year for public exposure to pathogens; return/increase of fish abundance/diversity; rate effects of different control alternatives; variance in the frequency and duration of SSOs; decrease or increase in basement backups; and change in operational safety factors.

Steps to estimate the values or benefits of various risk-management approaches include the following:

- Internal brainstorming by POTW to identify controls to minimize each wastewater system-related performance deficiency or failure,
- Screen the hypothesized controls to determine which are most effective, and
- Apply the refined risk measurement protocols to evaluate the screened controls.

2.5 Select a Target Level of Performance Risk

Step 5 is to select a target for performance for various system components. Examples include receiving waters free from wastewater-related pathogens during conditions expected in a typical weather year (defined based on data); specific aquatic diversity level; POTW rates within defined affordability limits; and number of backups per year.

Steps to selecting a performance target include the following:

- Develop and test means of comparing and balancing the alternatives;
- Facilitate stakeholder review of the alternatives;
- Refine the evaluations (which may require reiteration back through the risk performance measures); and
- Recommend target performance levels for each risk. Recognize that target levels may be defined as a time sequence, such as, achieve performance target X within 1 year, performance target Y within 5 years, and ultimately further progress toward performance target Z. This approach has been accepted in many CSO *long-term control plans (LTCPs)* and incorporates cost-effective analysis, break-point determinations, and risk scoring prioritization.

2.6 Adopt Objectives for Each System

Step 6 brings together the first steps to adopt performance objectives consistent with that target risk performance level identified. Steps to adopt these performance objectives include the following:

- Identify facility objectives consistent with each performance targets;
- Seek stakeholder acceptance of the performance objectives;
- Select final performance objectives and obtain regulatory approval (if required); and
- Develop an implementation plan with milestones, regularly monitor implementation, and report back to stakeholders.

This concludes the discussion of the six-step process, which was outlined in the introduction to Section 2.0 and discussed above in Sections 2.1 through 2.6. Systematically working through this process results in defined performance objectives and provides a structured approach to wet weather management that documents alternatives and relative benefits from various investments by the POTW. This documentation allows a community and stakeholders, including regulators, to develop a plan that meets engineering, regulatory, and political goals for the community. The process is further enhanced if it is grounded in the principles for wet weather management. Application of this approach, particularly the last step, is consistent with the approach being used by many communities developing CSO LTCPs.

This approach to setting performance objectives is potentially far more complex than the traditional approach of merely meeting the latest industry or regulatory criterion. It is recommended because the traditional approach has often failed to reflect the valid concerns of many stakeholders. In cases in which stakeholder concerns are readily discerned and noncontroversial, this approach is greatly simplified.

A small community in Wisconsin provides an example of how this approach can be applied in a small, resource-limited community. The community is a satellite community connected to the major metropolitan WRRF via a pumping station. In accordance with the six-step process,

1. Community leaders recognized that the residents wanted no more backups from the pumping station, no overflows at the pumping station, and continued treatment of wastewater at the WRRF.
2. The person responsible for the pumping station noted that high wet weather flows resulting from high rainfall-derived infiltration and inflow threatened all of these values.
3. The operator and leaders agreed that lowering the amount of rainfall getting to the sanitary sewer would be a good means of reducing the risks.
4. Contracted consultants recommended lateral and sewer inspections and replacement of those that admit rainwater.
5. Despite the recognition that any sealing would necessarily be imperfect and would decay with time, community leaders decided to lower the risk by sealing sewers to the greatest extent possible.
6. The community contracted for inspections of private and public sewers, cooperated with property owners to replace failed laterals, and contracted for sealing of public sewers.

Even if POTWs have limited flexibility to incorporate watershed priorities, community values, and conditions into their performance objectives because of a consent decree, state law, or previous commitments, they should reexamine the proscriptive provisions if the program does not meet expectations or if new information is available. The U.S. EPA CSO Control Policy, issued in early 1995, began a period during which many POTWs with CSOs negotiated consent orders requiring development of CSO Long-Term Control Plans. When the Long-Term Control Plan was completed, the facilities and other controls, as well as a detailed schedule, were incorporated into the consent order. Initially, regulators were reluctant to allow changes to these detailed programs. However, over time, the regulators have accepted evidence that changes are appropriate and agreed to modifications to consent decrees. In addition, the regulatory focus on SSOs has led to POTWs negotiating new consent decrees addressing additional compliance issues. Many of these consent decrees are available on U.S. EPA's Web site through the home page of the Office of Enforcement. The lesson from this enforcement history is that POTWs and their partners must be proactive to achieve the best results for their communities and stakeholders. The values-based risk management approach discussed above provides the structure for a proactive planning process and leads to strong supporting documentation to achieve regulatory approval, whether for wastewater systems or for integrated planning for all wet weather flows.

3.0 REFERENCES

American Public Works Association; American Society of Civil Engineers; National Association of Clean Water Agencies; Water Environment Federation (2010) *Core Attributes of Effectively Managed Wastewater Collection Systems;* National Association of Clean Water Agencies: Washington, D.C. http://www.wef.org/AWK/pages_cs.aspx?id=1063 (accessed May 2013).

American Water Works Association Research Foundation (2001) *Capital Planning Strategy Manual;* American Water Works Association Research Foundation: Denver, Colorado.

National Association of Clean Water Agencies (2002) *Managing Public Infrastructure Assets to Minimize Cost and Maximize Performance*; National Association of Clean Water Agencies: Washington, D.C.; Feb.

Swanson, G.; Kraus, T. (2009) Values-Based CSO LTCP Project Selection Process. *Proceedings of the Water Environment Federation Collection Systems Specialty Conference*; Louisville, Kentucky; Water Environment Federation: Alexandria, Virginia.

U.S. Environmental Protection Agency (2005) *Proposed EPA Policy on Permit Requirements for Peak Wet Weather Discharges from Wastewater Treatment Plants Serving Sanitary Sewer Collection Systems. Fed Regist.*, **70** (245), 76013.

U.S. Environmental Protection Agency (2012) Integrated Municipal Stormwater and Wastewater Planning Approach Framework; Office of Water and Office of Enforcement and Compliance Assurance; U.S. Environmental Protection Agency: Washington, D.C.; May.

Chapter 3

Guidance Practices

This chapter presents protocols for selecting wet weather management practices from the universe of traditional and evolving technologies available. The Water Environment Federation (WEF) gathered information from industry practitioners to identify these practices. The following hierarchy of terms and numbering system forms the basic structure of this chapter:

Major Process: x.0

Minor Process: x.x

System: x.x.x

Protocol Step: x.x.x.x

Protocol Substep: x.x.x.x.x

Practices

This chapter is organized by major process aligned with the major activities or functional areas in which a utility engages to accomplish its mission (planning, managing, and operating and maintaining), then by minor process, then by system (either conveyance or treatment), then by protocol step. For each protocol step, the guide presents principles applicable to the step (or substep) and then presents examples of how to apply the principles to select practices appropriate to specific cases.

1.0 PLANNING

Strategic planning is essential and must be done well to achieve stated organizational goals or requirements imposed by some other authority, such as the issuance of a National Pollutant Discharge Elimination System (NPDES) permit or enforcement order to a municipal utility. To be successful, strategic planning should be achieved efficiently, in a logical order, and reflect the organization's values. Without planning, the expended work will risk falling short or missing the desired goals. Adhering to a good strategic planning process will seek to develop an optimal solution that is cost-effective and meets the stated goals.

Strategic wet weather planning focuses on several important process steps. The process steps involve the following:

- Integrating conveyance and treatment in terms of flow interdependence and cost-effective alternatives analysis;
- Involving key staff, decision-makers, and stakeholders within the utility and stakeholders outside the utility in the planning process; and
- Selecting alternatives that have considered and represent the values and levels of service (LOS), including regulatory requirements, which the utility and community agree to accept once the work is completed and implemented.

The steps for strategic wet weather planning are illustrated in the Facilities Planning Flow Chart presented in Figure 2.3.

Many of the planning practices contributed by wastewater industry representatives for this document highlighted the importance of incorporating innovative technology and management tools and approaches to help solve complex problems. For instance, hydraulic modeling, optimization tools, and performance measurement processes like the balanced scorecard, stakeholder involvement, and risk-based prioritization help build decision confidence.

This guide presents the planning process into three minor processes, which are as follows:

1.1 Characterize Existing Conditions,

1.2 Determine Future Needs,

1.3 Select the Plan to Meet Those Needs.

Minor processes 1.1 and 1.2 represent the key initial steps of the planning process. The starting point, 1.1, addresses the existing system and its performance. Minor process 1.2 reflects the need to establish levels of service and performance objectives for the future as introduced and described in Chapter 2, Section 2.0. Chapter 2, Section 2.0 advocates a six-step, values-based risk management approach illustrated in Figure 2.4 to defining levels of service and performance objectives for management of wet weather effects. It provides guidance for applying risk and asset management techniques, in cooperation with publicly owned treatment works (POTWs) stakeholders, to identify community values to protect and associated acceptable LOS and performance standards.

To emphasize the importance of integrating conveyance and treatment systems, the discussion of minor processes 1.2 and 1.3 covers conveyance and treatment collectively as interdependent pieces of the utilities' overall wastewater system for delivering desired LOS.

Minor process 1.3 involved the compilation and analysis of data and information gathered in the preceding processes. It also includes steps illustrated in Figure 2.4 of the values-based risk management approach.

The contributed practices reveal that the planning process is a matter of educating, informing, and negotiating among stakeholders, and discussing technically achievable solutions and their financial and performance consequences. Often what appears to be the most technically and financially attractive solution is not always immediately recognized or even selected in a final plan. Factors, such as the community's values, political climate, legal constraints, or regulatory agency policy interpretations are some of the factors that can influence the selection of a final plan. An example of this is documented in *AMSA Wet Weather Survey, Final Report* (NACWA, 2003). One of the report's basic findings during the planning and design process was, "Technology exists to improve peak flow treatment performance and capacity, but the lack of regulatory compliance clarity and outright objections of regulatory authorities have prevented POTWs from installing the new technology."

Good planning practices for wet weather will bring together affected stakeholders to determine how they want to prioritize their limited resources among their values, such as LOS objectives, the environment, and public health concerns. These values and their priority will vary from community to community and from system to system. There is no certain path to compliance for wet weather plans.

Planners and designers must extend their thinking beyond design tables and standardized approaches. This point is further emphasized in the recommendations of *Financial Capability and Affordability in Wet Weather Negotiations* (NACWA, 2005). The paper recommended that, "Use of non-traditional and market-based approaches such as use attainability analyses (UAA's), watershed permitting, credit trading, phase implementation of requirements, and adaptive management will provide communities with the tools to ensure that maximum benefits can be achieved with affordable investments over time, to the net benefit of the communities served and the environment." Moreover, the release of the U.S. Environmental Protection Agency's (U.S. EPA's) "Integrated Municipal Stormwater and Wastewater Planning Approach Framework" (IPF) (2012) may offer opportunities to expand wet weather planning to include stormwater projects to the evaluation and prioritization of alternatives for wet weather controls.

The minor process guidance below will help stakeholders be aware of how others have approached the planning process and evaluated the particular challenges that wet weather flow management presents.

1.1 Characterize Existing Conditions

Characterization of existing conditions encompasses summarizing what is known about the existing service area, the flows generated, the infrastructure that collects and conveys the flows, and the facilities that treat and discharge the flows back to the natural environment. In addition to the principles presented in Chapter 2, the following principles are specific to selecting appropriate practices for characterizing and evaluating existing conditions:

A. Make maximum use of existing information sources. Part of the gathering of existing information should include interviews with operations and maintenance staff that have experience working on the systems. Valuable information that is not documented or not clearly recorded can be acquired by consulting with those who are intimately familiar with the system. Time spent with individuals who maintain a system can result in a better understanding of what the constraints are, where problem areas are located, and other information that could result in a better understanding of the system. In addition, review of customer compliance records is also a valuable source of information. Evaluating this information may help identify where chronic problems exist or provide insight to what actually occurs in the system during high weather flows. Not only will this information save time and money, it will encourage those involved in operations and customer relations activities to continue providing objective summaries for future use. It will also diminish the risk of failing to recognize limitations or system conditions known to others. Be cautious, however, to investigate and evaluate the quality of any data in consideration of how they will be used. The process of gathering information from existing sources provides an opportunity to evaluate data gaps that may exist along with how existing data are gathered and stored. Recommendations for improving data collection and storage should be captured for continuous improvement as a part of the process of gathering existing information.

B. Understand the performance evaluation process before recommending data collection or analyses. Carefully tailor the data collection to gather what is needed for the initial evaluation, measuring ongoing performance, and confirming program success in the future. Understanding critical control points in the system before collecting field data is necessary to optimize field data collection. A good plan for data collection will provide that limited resources are focused on gathering the critical data that are needed and that superfluous data collection and evaluation is avoided.

C. Use information storage and retrieval systems that facilitate data gathering, quality assurance, and the anticipated performance evaluation. Storage and retrieval systems must be convenient to each end user for access and use. Storing the information in the memory or notes of the sewer inspector may be convenient for that individual, but that approach may hinder the performance evaluation by making information inaccessible to a wider audience in the future. Storing the information in a sophisticated, secure computerized database often makes it inaccessible to all but a few trained data managers. Storing information in a manner accessible to analysts but not linked to quality assurance review may foster faulty data analyses. Maintaining data integrity as part of the information storage and retrieval systems is critical. These systems should help facilitate data analysis, thereby creating knowledge about the system and enabling the organization to make timely and educated decisions on the performance of the system. Many utilities develop data management planning documents at the beginning of system characterization activities that address how to sustain data management throughout the life cycle of system assets. Good resources that discuss data management systems in more detail are *Core Attributes of Effectively Managed Wastewater Collection Systems* (APWA et al., 2010); *Wastewater Collection Systems Management* (WEF, 2010); *Prevention and Control of Sewer System Overflows* (WEF, 2011); and *Innovative Internal Camera Inspection and Data Management for Effective Condition Assessment of Collection Systems* (U.S. EPA, 2010).

D. Use analysis methods or techniques that are consistent with the known conditions of the system. Make simplifying assumptions only when the evaluation process is known to be insensitive to possible errors introduced in the assumption. If such a determination is not clear or in question, perform appropriate sensitivity analyses and/or field investigations to test that assumption. Using analysis techniques that incorporate assumptions that may not be correct for wet weather conditions may result in faulty performance evaluations (e.g., using analysis techniques that assume the hydraulic grade line is parallel to the pipe grade line will overestimate conveyance capacity in a system subject to backwater from pumping station wet wells or downstream surcharge).

E. Use practices consistent with local practices that have been successful in the past unless they violate other overriding principles. Consistency with past work facilitates trend analyses and greatly expedites understanding and completing the performance evaluation.

Protocols and the practices typical of characterizing existing conditions are summarized in the following subsections, first for conveyance and subsequently for treatment systems.

1.1.1 *Conveyance*

The characterization of existing conditions involves summarizing what is currently known about the sewer system, gathering important missing information, completing analyses, and documenting the performance characteristics of the existing system. The protocol steps for minor protocol Characterizing Existing Conditions in a conveyance system involve:

- Defining the sewer service area, which is the geographic area that encompasses the wastewater system;
- Quantifying existing flows, including base sanitary flow, groundwater infiltration, and rainfall derived infiltration and inflow (RDII);

- Determining component and system capacity, including the existence and behavior of hydraulic surcharging; and characterizing the conveyance network, including the location, elevation, size, age, and construction characteristics including material type; and
- Defining the capacity status while recognizing original performance objectives, existing limitations, capacity constraints, regulatory performance requirements, frequent clogging, frequent complaints, and the extent to which other systems (such as rivers and storm sewers) influence system performance.

Commonality exists among conveyance systems such that protocols and practices may be considered as standard procedures; however, not all systems and resources are the same. Applying these protocols will result in selecting differing practices for differing collection systems.

Consider the example of a small collection system served by a package POTW that was installed by a subdivision developer who did not keep detailed construction records. The most appropriate practices may include reviewing monthly discharge reports to determine if the POTW shows a marked wet weather response, and interviewing the part-time operator and sewershed residents to determine if they have observed any wet weather problems. If the POTW records, the operator, and residents reveal no evidence of sewer system problems or wet weather response, it may be appropriate to conclude the performance evaluation by simply documenting that the available information is consistent with the assumption that the conveyance system performs adequately, even during wet weather. Emphasizing the first and second characterization principles listed above can quickly lead to simple practices that meet the listed principles.

On the other extreme, a collection system may be a moderately sized network built over several decades and is tributary to a water resource recovery facility (WRRF) managed by a large utility in a major metropolitan area. Existing information may likely be far more extensive and include extensive evidence of wet weather problems or past analyses of system performance and ongoing performance monitoring. In this case, the final performance evaluation may include using dynamic hydraulic modeling software capable of estimating overflows and/or basement backups in unmonitored locations. The technique appropriate to developing this analysis using computer simulation tools would be largely dependent on preexisting tools and information. If the community already has base mapping in a geographic information system (GIS), it would be wise to link information storage, retrieval, and analysis to that system. However, if the community does not have a GIS, or has existing information stored in isolated systems, processes demanding GIS linkage might violate the characterization principles A and E. Similarly, if the community has an existing hydraulic simulation model, but that model assumes that peak flows are independent of rainfall, whereas the flow records show marked rainfall response, use of that model would violate characterization principle D.

Approaches for applying the listed principles to select appropriate practices in a given situation are described below as four distinct protocol steps:

- Define Service Areas,
- Quantify Flows,
- Determine Component and System Capacity, and
- Define Capacity Status.

1.1.1.1 Define Service Areas

Defining service areas for wet weather wastewater planning may range from simply identifying the geographic area served by the existing POTW and satellite systems to subdividing service areas into specific conveyance network tributary areas often called "sewersheds" when associated with sanitary or combined sewer systems. If doing integrating planning, then "catchments" may include storm sewers, depending on the naming conventions of the local utility.

Service areas may be characterized and spatially defined by the types of service and their physical similarities, such as separate storm sewers, separate sanitary sewers, separated sewers, combined sewers, and further subdivided by their attributes if desirable. The characterization principles previously articulated (Section 1.1) should be used to define what level of detail is required, and then practices should be selected that supply the required level of resolution.

In characterizing the existing conveyance system, the key principle to apply is to understand the performance evaluation process before recommending data collection or analyses. For example, if one must prepare a combined sewer overflow (CSO) long-term control plan for a multicommunity system, then a more detailed characterization would be required. Guidance for characterizing such a complex system was previously published in guidance documents for long-term control planning (U.S. EPA, 1995a; 1999).

1.1.1.2 Quantify Flows

Three key questions must be addressed in quantifying the flows.

- How spatially discrete (i.e., at what scale) must the quantification be?
- What flow conditions must be quantified?
- How accurate must the quantification be?

In a wet weather analysis, the spatial delineation must be sufficient to account for all flows into the system, even those that do not reach the WRRF, and in a spatial resolution that accounts for the service area definitions performed in the preceding step. Consequently, it is typically necessary to use flow quantification techniques that account for overflow quantities, including unidentified discharges. Unidentified discharges commonly include those in which the discharge is submerged, and flows unobserved into waterways (e.g., backflows through catch basins into gutters in combined sewer areas, unreported discharge pumped from manholes to protect basements during significant storms).

In a wet weather analysis, the conditions quantified must include peak wet weather conditions. This is complicated by the fact that most systems experience backwater and surcharge during peak wet weather, whereas most flow-quantification techniques are based on an assumption of "normal" flow conditions in which the sewers are flowing with no surcharge and no backwater. Furthermore, for analysis of sewer system performance during peak wet weather conditions, it is necessary to identify peak flows for short durations, that is, periods of duration consistent with the time of concentration for each portion of the sewer network, which may be as short as 5 minutes in some cases. Supporting this flow monitoring and subsequent analysis should be an equivalent level of rainfall monitoring at a data collection frequency consistent with that used for flow monitoring.

Given the above complications during wet weather, the question of accuracy becomes one of how accurate can the quantification be. The accuracy attainable under ideal normal sewer flow conditions cannot be necessarily expected under all wet weather conditions, particularly because the opportunities for calibrating and testing the flow quantification methodologies seldom coincide with peak wet weather conditions.

The preferred method for quantifying flows in an existing system is flow monitoring. Numerous publications have addressed questions of where and how to monitor, and several manufacturers of flow monitoring devices provide training as well as primary devices for flow monitoring. It is necessary to review the principles of characterizing existing conditions (minor process 1.1) and the literature and adapt monitoring methods to the defined site-specific wet weather conditions. If appropriate for monitoring the long-term performance of the system, one should consider making certain strategic monitoring locations permanent, including a dedicated connection with a supervisory control and data acquisition (SCADA) system.

Wet weather capacity assessments of collection systems are necessarily based on flow quantification, but flow monitoring during peak wet weather conditions may be incomplete and inaccurate. The limitations of peak wet weather flow monitoring are necessarily overcome with assumptions and calculations (often embodied in models). Each assumption incorporated in the calculations should be identified, evaluated, and documented so that future users of the quantification can adapt it appropriately. Each use of the flow quantification should explicitly address the limitations inherent in the original flow quantification and draw conclusions that are appropriately qualified.

The analyses must always recognize that "the flow" is, in reality, an estimate of the flow rate under only one of an infinite variety of weather conditions (rain intensity, volume, duration, antecedent precipitation/moisture conditions). It is becoming more of a standard practice to evaluate system flow performance under a variety of peak flow conditions, because experience has shown that systems perform differently in short-term (typically high-intensity) versus long-term (typically higher total depth) rain events. The statistics of storm and flood probability are useful tools in addressing the variability of sewer flow rates. Rather than defining "the peak flow", it is recommended that the flow quantification address "the peak flow with a return probability of. . .". Hydrology textbooks describe the calculation of the return probability. Note that the hydrologic theory will demonstrate that the probability of a peak storm flow is seldom the same as the probability of the associated rainfall intensity or total depth.

1.1.1.3 Determine Component and System Capacity

The wastewater conveyance system is commonly composed of gravity sewers, inverted siphons, pumping stations and force mains, storage systems, and flow control devices. Each of these components has an individual capacity. The system is the network of multiple individual reaches of sewers, pumping stations, and other features that work together to convey flow to a downstream point. The conveyance system capacity is determined by initial design criteria, installation techniques and accuracies, the extent to which the system has been maintained, and how these individual component capacities interrelate.

1.1.1.3.1 Gravity sewers

Gravity sewer capacity is often defined as the amount of flow a sewer of a certain size and slope is designed to carry. Design guides typically call for sizing sewers to flow at a depth less than full normally during dry weather conditions. Hence, the related design calculations typically include an inherent assumption of hydraulically normal flow, that is, conditions in which the water surface through the pipe is parallel to the slope of the pipe, as defined by the upstream and downstream inverts. Those conditions, illustrated in Figure 3.1, are conditions in which the Manning's equation can be applied using the pipe slope.

For this hydraulically normal condition (commonly referred to as "normal depth"), the capacity of each reach of sewer is assumed to be the amount of flow that can be carried in the sewer without filling (surcharging) the sewer pipe. There are systems that were originally designed to flow with surcharged pipes, and in those cases, the design capacity was not intended for the hydraulically normal condition.

During wet weather conditions, many existing conveyance systems experience conditions in which they carry more flow than their design capacity. If flows greater than the design capacity enter the sewer, the hydraulic grade line slope

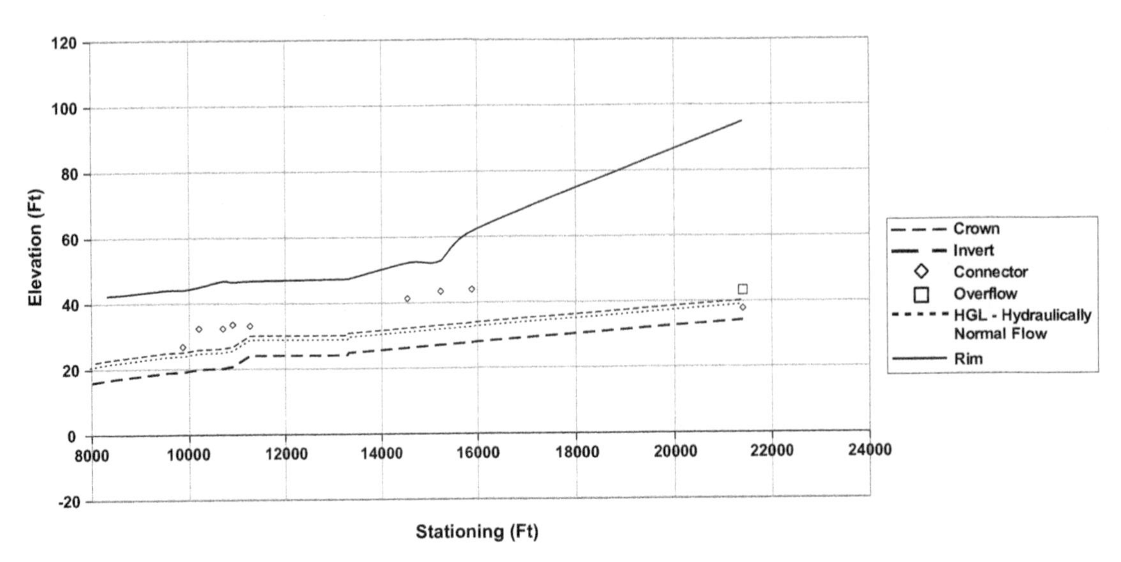

FIGURE 3.1 Sewer profile under hydraulically normal flow (HGL = hydraulic grade line; ft × 0.3048 = m).

may increase until the constrained flow finds a relief point (i.e., an overflow, a basement to back into, or the ground surface). In some circumstances, the elevated water surface can force more flow through the downstream sewer reach than the calculated capacity.

If the "hydraulically overloaded" sewer is in a downstream portion of the conveyance system, then the upstream sewers find their discharge point "surcharged". To convey flow downstream, the water level at the upstream sewer outfall must rise above the surcharge elevation of the downstream sewer, and pipes upstream experience a similar rise in water level as illustrated in Figure 3.2. In this case, the upstream sewers are carrying flows well within their calculated design capacity, but they are nonetheless backing up and experiencing overflows. Overflow to another pipe will occur where the hydraulic grade line exceeds the overflow elevation (right end of the diagram). The connectors shown in the profile could be either additional municipal sewers or individual property sewer laterals. If the connectors are sewer laterals, surcharging the connector may cause a basement backup. If the connectors are municipal sewers, surcharging at this location could affect their effective capacity. This type of graph can be used to identify those sewers that are being constrained by downstream conditions as those in which the water surface profile slope is less steep than the pipe slope.

Thus, individual sewer reaches within a conveyance system can carry more than their design capacity with no adverse consequences, while individual sewers within the conveyance system can back up, overflow, and/or flood even while carrying flows well within their design capacity. Consequently, it is necessary to

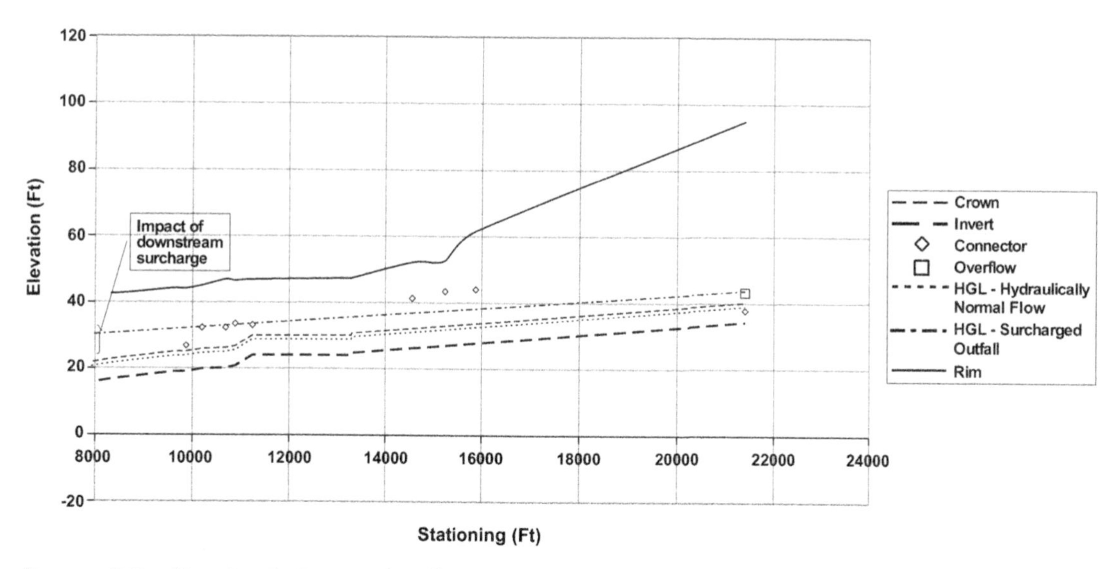

FIGURE 3.2 Overloaded sewer's effect on upstream pipes (HGL = hydraulic grade line; ft × 0.3048 = m).

define a system capacity independent of the design capacity of the individual reaches of sewer.

1.1.1.3.2 Inverted siphons

Inverted siphons (sometimes just called "siphons") are features in conveyance systems that allow a gravity sewer to avoid an elevation conflict (such as another pipe or a river) in its downhill path toward a downstream point. Siphon capacity is typically defined as the amount of flow through the system when the upstream driving head is at the upstream sewer design flow depth (e.g., 80 to 90% full) and the discharge sewer is similarly 80 to 90% full. The siphon capacity calculation necessarily accounts for friction losses, and often accounts for entrance, exit, and transition losses. Like sewers, siphons cannot carry their calculated design flow if the discharge is surcharged, and they will carry more than their design flow if the upstream sewer is surcharged.

Many communities require multiple pipes, or "barrels", at siphons, and define a firm capacity of the siphon assuming that one barrel is out of service. Siphons are locations where overflows can occur and should be assessed for wet weather performance characteristics, including the potential for accumulating silt deposits. It is important to understand the actual maintenance practices related to inverted siphons, such as sediment flushing, so the effective capacity of these features is not overstated.

1.1.1.3.3 Pumping stations and force mains

A pumping station is, in itself, a system that may be composed of a wet well, headworks (including bar screens or grit collection boxes), pump or pumps, pump controls, and discharge force mains. A particular type of pumping station, termed a "lift station", lifts wastewater from a lower sewer into a higher sewer to allow flow to continue by gravity. The capacity of a pumping station is determined in terms of flow and amount of pressure head the pumps must overcome. The pressure head, or total dynamic head (TDH), has two components: the static head that is related to the elevation difference from the pump intake to the discharge point or highest point in the force main, and the dynamic head generated by the friction in the discharge force main. The dynamic head is a function of the flow rate and the size, roughness and length of the force main pipe, inlet conditions to the pumps, minor losses, and the configuration of the downstream force main system. The flow passing through an individual pump is defined by a pump curve in which the efficiency of the pump varies with the TDH of the discharge. Stated simply, the amount of flow that can go through a pumping station is a complex interaction of the machinery, the force main, and the head in the receiving sewer.

Pumping station capacity is often defined in terms of both "firm capacity", the amount of flow that can be passed through the station with one (typically the largest) or more pumps out of service; and the "peak capacity", the amount of flow that can pass through the station with all pumps operating at capacity. Pumping station capacity, firm and peak, varies as the efficiency of the pumps degrades with time. The analyst should determine what station capacity can be used for simulating the systems wet weather performance. Appropriately experienced engineers should evaluate in detail all situations involving combinations of pumps, manifolded force mains, and other advanced system controls.

In wet weather, pumping station controls are used to quickly switch the station from normal flow to peak flow mode. Pumping stations are critical and often complex components of the sewer system. Assessing system flow characteristics in wet weather against system capacity must consider station controls, valving, wet well storage, and regulatory standards. Pumping stations are also likely locations for wet weather overflows in older systems.

1.1.1.3.4 Flow control devices and storage facilities

Additional flow conveyance management features that can be present in some systems and are important in defining capacity include flow control devices and storage facilities. These features are particularly important in the context of this guidance document as they are routinely used in wet weather flow management.

Flow control devices is a general term used for any variety of infrastructure components that throttle flow in a sewer to actively store water upstream or divert flow from one pipe to another. These components may have static settings for when they are in use, or may be dynamically controlled as a result of localized conditions or the intervention of a sewer system operator. Gates, vortex flow controls, and weirs are common types of flow control devices.

Storage facilities are typically installed at locations that are chronic sewer system relief points or overflows. These facilities are used to reduce the frequency and amount of discharge to the natural environment. Storage facilities have a designed amount of capacity in terms of storage volume. This can be challenging to compare to other "flow-through" types of capacities that have been discussed for features such as sewers and pumping stations. If storage facilities are present in a system, it may necessitate using more advanced tools, such as dynamic hydraulic models, for defining system capacity status. Additional means of providing storage include flow control devices in concert with sewers to produce "in-line" storage of peak flows.

1.1.1.3.5 Conveyance system capacity

Defining conveyance system capacity requires definition of the capacity to be used for each component and definition of how the various components interact.

The term "acceptable system surcharge capacity" has been used in complex systems. This term is defined as the amount of flow that can be carried in the conveyance system without causing property or infrastructure damage or overflow. For peak wet weather conditions, the capacity is often calculated twice: once assuming that all siphon barrels and all pumps are active, and again assuming some siphon barrels and some pumps are not active. In some systems, any amount of surcharge is considered damaging, in which case acceptable system surcharge capacity would allow no surcharge.

In any case, calculation of acceptable system surcharge capacity for a given conveyance system requires consideration of all of the sewer component interconnections in the system, particularly the downstream sewers that intercept flows from multiple areas. The condition assumed to occur at the downstream end of an analyzed conveyance system is commonly referred to as the "boundary condition". The common approach of assuming a free discharge boundary condition to calculate upstream system capacity should be used only in those rare cases in which that assumption is realistic during peak wet weather conditions. The more advanced sewer analysis tools and models allow for additional options for evaluating the effect of boundary conditions on conveyance system capacity.

Approaches to calculating the acceptable system surcharge capacity vary with the complexity of the system being evaluated. For simple systems, manual calculations using the Manning's formula are adequate if the analyst starts the calculation from the water surface elevation at the downstream outfall, and uses the slope of the water surface, not the slope of the pipe, in calculating the flows. For more complex systems, computer models that address surcharged flow conditions are appropriate. Before initiating an analysis of system capacity, it is important to select an analytical method that considers the presence of each of the features discussed in the above section.

1.1.1.4 Define Capacity Status

The capacity status of the conveyance system answers the question: Does the conveyance system have adequate capacity to carry the flows it receives for the performance target? The performance target would typically equal the original design capacity; however, the performance target may have been later revised to address level of service objectives and, as a result, be greater than the original design capacity, particularly if the design capacity did not include a wet weather flow allowance. Also, as mentioned above, the conveyance system may convey more flow than the design capacity without an overflow if, for instance, surcharge was not included in the original design capacity.

Intuitively, definition of capacity status requires comparing the performance target flow to existing conveyance system capacities. It should be noted that different systems may have significantly different performance targets related to overflow control. For instance, CSOs to an industrial shipping canal may have a lower target, but CSOs near a sensitive area such as a public beach may have a higher target, and sanitary sewer overflows (SSOs) may have an even higher target.

Another approach to defining conveyance system capacity status is to assume that if the conveyance system is never observed to overflow, then it has adequate capacity, and if it has been observed to overflow, then the capacity is inadequate. In the first case, the assumption would be valid only to the extent that there are no unobserved (submerged or remote) overflow locations or connectors adversely affected but not overflowing. In the second case, overflows could be the result of factors other than inadequate capacity, such as blockages, downstream surcharges, failed pumps, and power failures.

The simplest approach to defining capacity status is to compare recent monitored peak flows to the capacity of the components nearest to those monitored flows. Even in simple systems, this approach requires adaptations that relate flows at discrete monitoring locations to flows at specific components that may be far away from the monitoring point. This approach can be summarized in a tabulation of system components, component capacities, and peak flow near each component. This approach has limitations if the recent monitored data only includes events that are less demanding than the performance target established for the system.

For complex systems, two options are available: (1) define capacity status in terms of performance records or (2) determine capacity status using complex system simulations. In both complex cases, the principle of maximizing the use of available data requires that the available information (e.g., WRRF flow records, pumping station operation records, system flow monitoring, overflow activation records, property sewer backup complaint records) be compiled and analyzed. If the initial compilation provides information that reflects adequate systemwide performance under wet weather conditions similar to or greater than the performance target, then the capacity status can be completed by organizing and documenting the findings. Regulators often use this approach to determine if specific compliance steps are necessary for a system.

Using performance records to define capacity status works well in those systems in which existing performance records reflect performance of most of the system components under peak wet weather conditions. Take, for example, systems that are instrumented such that flow depth is continually recorded throughout the interceptors, several connections are instrumented to record depth and estimate flow, and all known overflow points are instrumented to detect activation and estimate flow. One utility in this group might use these performance records to document which wet weather conditions exceeded capacities in various components of the system. Another utility might operate permanent flow meters connected to a SCADA system that reports real-time data that are used by system operations and management. Still others might make extensive use of temporary flow meters, varying the location periodically to cover multiple sewersheds. In each of these cases, it is important for the system owner that relies on data alone to include in the records the types of wet weather events that are similar to or greater than the performance target for the system.

Most system managers, however, find that the available data do not reflect systemwide conditions during peak wet weather conditions in events that approximate the performance target or that conditions other than those that were directly measured require analysis. In these cases, simulation models are used to extrapolate, or in some cases interpolate, the limited available data to locations and weather conditions not represented in the data. Despite these limitations, the existing data are still used to calibrate and verify the simulation tools.

The utilized simulation tool can be complex or simplistic, depending on the system to be modeled and data used to calibrate it. If WRRF flow records are the only wet weather data available, one could develop a simple hydrologic relationship between rainfall and flow and determine whether the relationship is consistent with a variety of rainfall conditions. If increasing rainfall and/or intensity do not result in increasing flows, the data could indicate upstream hydraulic constraints and overflows, while an increasingly steep relationship might indicate inflow from cross-connections or manholes affected by surface flooding. These initial indications could be verified against complaint records and operator observations.

If initial review of existing data indicates a significant lack of capacity to handle peak wet weather conditions, or if anticipated growth will demand more capacity, it is advisable to move toward more complex rainfall distribution and system simulation models that can later be used in alternative evaluation and planning. The choice of the model should be based on the characteristics of the system, the intended modeling approach, and the availability of data to support model development and calibration. Any simulation tool can be applied with "assumed" data (to be replaced later with better data as it becomes available), but simulation tools that make simplifying assumptions can never analyze complex problems. For example, a steady state model that is incapable of simulating changing flow conditions will never be able to evaluate properly a system that uses storage facilities to manage peak flows. Although the calibration of these models can require a substantial amount of data that can be expensive to acquire, having sufficient data to adequately define the problem at the performance target can help avoid more expensive capital investment mistakes.

1.1.2 *Treatment*

Characterization of the WRRF involves summarizing what is currently known about the operating conditions, practices, and facilities within the WRRF, gathering missing information, completing analyses, and documenting the performance characteristics of the existing system.

The protocols for characterizing existing conditions at a WRRF include the following:

- Quantify and accurately measure flows and wasteloads;
- Summarize the characteristics of the service area; and
- Determine the capacity of the WRRF system and individual components, including redundancy capabilities.

The Clean Water Act (CWA) requires that, at a minimum, publicly owned WRRFs meet secondary treatment standards. During exceptional circumstances, a "no feasible alternatives" analysis, which is included in discharge permit conditions, may allow a portion of the wet weather flow in a WRRF treating combined wastewater to receive only primary treatment (or its equivalent). Depending on receiving water characteristics, enhanced treatment standards may be applied in addition to secondary standards. As a result of the variability in unit processes used at the WRRF and the uniqueness of the flow conditions and wasteloads, not all WRRFs perform in the same manner. Their ability to handle increases in flow and loading vary depending on the variability of these flows and loads, the applicable local design standards and regulatory requirements, preferences of the designer, and the levels of operation and maintenance performed at the facility.

Detailed historical operating data for smaller WRRFs may be less readily available than for larger facilities, but that is not always the case. Accessibility of computerized data management systems and enhanced instrumentation at all WRRFs can provide a wealth of historical data for analysis. Searching the precipitation record for especially wet periods can be valuable to evaluate collection system response to antecedent moisture conditions. For communities adjacent to large rivers, evaluation of river stage independent of precipitation is important for evaluation of flow volume at the WRRF. It often comes down to the resources the utility applies to operation and maintenance of the facilities and the curiosity and organization of the staff at the WRRF.

Recent facilities plans often have good information available for use in characterizing the existing conditions. Evaluations that have been performed of the conveyance system should also be reviewed to determine the potential magnitude of wet weather flows and their effects at the WRRF.

1.1.2.1 *Quantify Flows and Loads*

The following two key questions must be addressed in quantifying flows and loads received at the WRRF:

- What conditions must be quantified?
- How accurate must the quantification be?

The intensity, duration, and frequency of the wet weather event and its resulting hydrograph (variation in flow with time) and pollutograph (variation in a pollutant's load with time) at the influent to the facility are the important conditions that must be quantified in this analysis. Consideration should also be given to spatial and directional distribution of rain events in a particular area. Historical data offer some insight (in particular with regards to flow), but future conditions, such as expansion of the service area, types of customers, and maintenance level of the conveyance system, must be considered in estimating future peak wet weather conditions.

In a wet weather analysis, the conditions to be quantified (peak month, peak week, peak day, peak instantaneous) could include flow rates well in excess of those typically experience by the WRRF during dry weather operation. An estimation of some of these peak wet weather flow conditions may be needed for WRRFs that have difficulty accurately measuring flows over a wide range.

Development of pollutographs requires the collection of discrete WRRF influent samples at frequent intervals throughout the event, and in particular during the first events of the wet weather season when the highest first-flush conditions (a rapid change in pollutant loadings due primarily to the resuspension of solids that have settled in the conveyance systems during low flow conditions) are to be expected. This is not commonly practiced at WRRFs, thus often requiring dedicated sampling programs to generate the data. Because each storm generates a unique hydrograph and pollutograph, several storms should be characterized over a range of seasons. The intent is to develop an understanding of the service area's response to storm events and using that understanding to establish design parameters for wet weather treatment.

Whereas the variability of storm events precludes precise predictions, quantifying wet weather event conditions is important to the design and operation of the WRRF. Depending on the overall capacity of the conveyance system and the WRRF and its type and number of individual unit processes, a relatively small error in estimating the corresponding hydrograph and pollutograph could make it difficult to handle appropriately these conditions at the WRRF. For example, it is not just the extent of the higher flow rate, but also the duration of this extraneous operational condition that will determine the ability of a WRRF to maintain compliance with its effluent requirements. Short-term spikes in flow or load can often be dampened in the treatment process if appropriate provisions are made in the design. Longer term, flow, and load conditions outside of established design parameters (including prolonged periods of high flow and low organic loads typical of the late stages of wet weather events) may result in hydraulic or treatment performance issues. Investing the resources to accurately estimate flow and load influent conditions during wet weather events is typically a worthwhile investment.

The peak wet weather conditions experienced at the headworks of a WRRF is a function of the flows and loads that can be delivered by the conveyance system typically under surcharged system conditions. The nature of those flows and loads in terms of duration and intensity are also dependent on the characteristics of the conveyance system. The discussion in the previous section on conveyance should be reviewed to characterize the wet weather conditions that could be experienced at the WRRF.

Unit processes at the WRRF downstream of the headworks will influence the effect of wet weather conditions on subsequent unit processes. Dynamic hydraulic and process simulation (considering not steady-state but rather time-variable input parameters) will help enhance understanding of the effects on peak flow and load dampening and overall pollutant removal within the WRRF.

The use of raw wastewater pumping either in the conveyance system or at the WRRF has a large effect on how the WRRF will need to respond to wet weather conditions. Allowing some storage in the conveyance system piping, pumping station wet wells, and variable-rate pump control systems provide some attenuation of peak flows and may give the WRRF staff some control of the flow rate delivered to the WRRF. Wet weather flow equalization basins at pumping stations can further attenuate peak flows and loads to the WRRF. Gravity sewers can deliver widely varying quantities of wastewater to the WRRF, and can deliver large amounts of large-sized trash and seasonal loads of leaves that could create materials handling problems at the WRRF.

1.1.2.2 Characterize Service Area

For most conveyance systems (both separate sanitary and combined systems), once first-flush conditions have passed, additional peak wet weather flows do not significantly increase the pollutant loading to the WRRF; it is assumed that most of the infiltration/inflow (I/I) does not bring with it a significant pollutant load. Figure 3.3 illustrates how flow and pollutant load influent to the WRRF might vary in a wet weather event. First-flush conditions can vary in intensity and duration depending on the size and configuration of the conveyance system, the rainfall intensity and duration across the service area, the

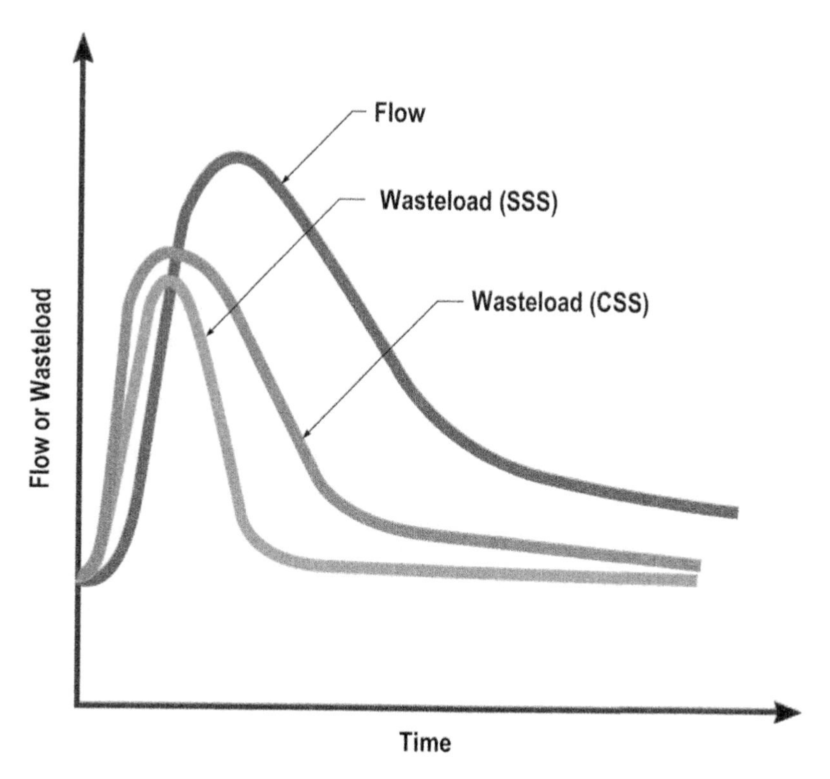

FIGURE 3.3 Flow and pollutant load response during a wet weather event (SSS = separate sewer system; CSS = combined sewer system).

antecedent rainfall and time of year, and the time interval since the last significant wet weather event. Combined sewer systems can experience longer durations of higher inorganic solids pollutant loadings to the WRRF during peak wet weather conditions than sanitary sewer systems, depending on the nature of the inflows. Increased levels of trace metals such as zinc and lead can also increase in combined systems resulting from the effect of pollutants from street runoff. Conversely, separated systems may experience a longer duration of high flow associated with infiltration until groundwater levels drop below the collection system. The discussion in the preceding section on conveyance service areas should be reviewed to determine how the WRRF influent would change with increasing wet weather flows.

1.1.2.3 Determine System and Component Capacity

A WRRF has many individual components that are combined to form a treatment system, including pumping systems and gravity conduits; physical, chemical, and biological WRRFs; and solids handling facilities. The capacity of the various components and the entire WRRF can vary with how the capacities are defined and how these systems interrelate. Hydraulic throughput and treatment capacities (both installed and firm) are defined below and are important to consider in discussion of capacities.

- Peak (hydraulic) capacity is defined as the instantaneous wastewater flow rate that can be passed through the WRRF with all channels and process units in service without resulting in equipment damage or overtopping of structures. Some WRRFs have the ability to shut off aeration of basins and channels allowing solids to settle during extreme wet weather events to convey higher flows and retain some solids at the bottom of the basins and channels. This has the dual benefit of reducing the potential for discharge permit violations caused by solids washout, and also retains active biomass in the system to facilitate restoring the biological process to normal operation following the rain event.
- Treatment capacity is defined as the maximum wastewater flow that can be passed through the unit process with all units in service without resultant failure of the treatment process to perform its intended function. This can be a variable value for biological systems depending upon the settleability of the biomass and the duration of the event.
- Firm capacity is defined as the maximum hydraulic or treatment capacity with the largest unit of each unit process out of service. This is used during design to ensure that a redundant unit is provided for critical unit processes.
- Gravity conduit capacity is defined as the amount of flow that a pipe has been designed to carry at specified upstream and downstream water surface conditions. Some conduits are designed to flow with surcharged conditions and some are designed to flow without being surcharged. The capacity of the conduits is often tied to the size needed to maintain a minimum scour velocity at low WRRF flows, thus setting a maximum capacity. A WRRF's hydraulic profiles often depend on the elevation of the receiving water. Peak WRRF flows and high receiving water levels may not always coincide, but worst-case scenarios of both should be considered during design because these conditions can limit hydraulic capacity of both gravity and pumping systems. During wet weather events, high flows can cause gravity conduits and control structures to be overtopped or cause the upstream water surface elevation to be greater than allowable, resulting in localized flooding. This is often an undesirable condition that can lead to damage to WRRF facilities, hazardous operating conditions, or poor process operation. Consequently, it is necessary to define system capacity based on the capacity of each gravity conduit and the effects of receiving water levels.
- Pumping system capacity is determined in terms of flow rate and the amount of pressure head the pump(s) must overcome (see Section 1.1.1.3.3). The amount of flow that can go through a pumping station is a complex

interaction of the machinery, the force main, and the head in the pumping station wet well and discharge point that is best described in other technical references.

- Physical, chemical, and biological treatment system capacities are often driven by regulatory design guidelines. Such guidelines can be found in individual state regulations, in industry standards, such as the *Recommended Standards for Wastewater Facilities* that is also known as the "Ten States Standards" (GLUMRB, 2004) and *Design of Municipal Wastewater Treatment Plants* (WEF et al., 2010) or practices of individual consulting engineering design firms. Capacities are often referenced in terms of an average condition and peak conditions. Experience has often led to some conservatism in the determination of design capacities, but the degree of conservatism will vary by the particular design standard. In practice, determination of the peak capacity for a biological treatment system is difficult, given the changing nature of flows, loads and duration related to wet weather events, and the consideration of the potential variability in biomass settling characteristics. Higher flow rates may be treated adequately if the flow is increased slowly and sufficient parameters are optimized before the wet weather event. Rapid increases in flow may cause performance to degrade before reaching the nominal peak capacity.

Each unit process and hydraulic component should be evaluated as to how it compares to its nominal wet weather treatment capacity. That capacity is typically documented by the designer during facilities planning, preliminary engineering, or final design of the WRRF. Capacities can often be found in the operations manuals prepared for the WRRF. Once the projected wet weather flows and loads are known and assessed against the capacities of the facilities, those conditions can be compared to current design standards or industry standards of practice. However, it should be noted that it is not uncommon that these rated capacities and standards are established on the basis of fairly constant, quasi steady-state influent conditions, which makes predicting performance under the very dynamic conditions of wet weather events more difficult.

The simplest approach to defining capacity status is to compare documented unit process and systemwide performance under actual wet weather event conditions. This approach can be summarized in a tabulation of unit processes, unit process capacities, and peak flow and load experienced at each unit process. If this information is not available, the best assessment of capacity, but one that can be costly and time consuming, is to perform hydraulic stress testing of the components under various field conditions to determine the flow conditions under which performance is unacceptable. For complex systems, two options are available: (1) define capacity status in terms of performance records or (2) determine capacity status using process simulations.

Stress testing can be an element used to determine capacities for both options. Stress testing guidelines and detailed protocols are available from a number of references, some of which are identified in the references section of this guide.

If operating data are available, the principle of maximizing the use of available data requires that the available information—WRRF and individual unit-process flow and performance records—be compiled and analyzed. If the initial compilation provides information that reflects performance under intense wet weather and low temperature conditions, then the capacity status can be completed by organizing and documenting the findings. Using performance records to define capacity status works well in systems in which existing records reflect performance of the unit processes under peak wet weather conditions.

For systems in which the available data do not reflect systemwide conditions and/or do not reflect performance during peak wet weather conditions, simulation models and/or stress testing are used to extrapolate limited available data to weather conditions not reflected in the data. The existing data can be used to calibrate and verify the simulation tools or additional data can be collected for that verification of the models.

The simulation tools can be complex or simplistic, depending on the complexity of the treatment system and the availability of existing data to reflect that system. Spreadsheet models as well as complicated dynamic models are available to predict WRRF unit process performance under a variety of conditions. Complexity of the model used can also depend on how the operators plan to use the results in control of the WRRF under peak wet weather conditions. Complicated control schemes (such as those typical of biological nutrient removal processes) can require a more sophisticated model to understand the effects of the decisions. The user needs to be mindful that the results of these simulations depend largely on the quality of data input and how well the model represents the actual operation.

If an initial review of existing data indicates a serious lack of capacity to handle peak wet weather conditions, it is advisable to move toward more complex dynamic simulation models that can later be used in alternative evaluation and planning. The choice of the model should be based on the characteristics of the system. Any simulation tool can be applied with "assumed" data (to be replaced later with better data as it becomes available), but simulation tools that make simplifying assumptions can never analyze complex problems.

1.2 Determine Future Needs

Planning for the future needs of a system should consider the following issues at a minimum:

- Anticipated changes in base wastewater flows from the service area from growth and redevelopment, including consideration of existing and expected economic conditions, as well as predicted changes in I/I that may be expected to occur from further system deterioration or planned rehabilitation;

- System sustainability issues, such as structural integrity degradation, operations and maintenance requirements, and the potential effects of future changes in wet weather flows (including I/I rates); and
- Regulatory requirements, including current requirements and anticipated future requirements.

All three of these planning considerations are interrelated and continually changing for most utilities. Changes in one of these considerations may influence the best methods of addressing the others, whether these are through rehabilitation and replacement program, modified operations and maintenance practices, or other capital improvements. Development of a wastewater facility (or wastewater system) plan is the typical process by which these issues are considered and decisions are made as to how to address these issues under future conditions.

Wastewater system planning involves establishing performance objectives, projecting future flows, identifying and evaluating alternatives, selecting the most appropriate alternatives, developing an implementation plan and schedule, and securing funding and commitment to implement the plan. Each step in the planning process must proceed with an eye toward achieving the established performance objectives in the system.

1.2.1 Estimate Future Flows

Future flow projections must identify how flow will change with growth, redevelopment, and changes in the management of wet weather flow sources. It is essential to incorporate consistent evaluations of future flows throughout all wastewater system elements. Future flows should consider existing economic conditions and how those conditions may change in the future. Planning and design assumptions for conveyance, treatment, and potential storage of wastewater flows should be conducted such that the implemented facilities will operate in a complementary manner. This should include consideration of facility phasing, validation of expected flow changes, and system operations through flow monitoring.

1.2.1.1 Planning Period

Construction grants requirements during the 1970s encouraged the preparation of facilities plans for a 20-year period. Nevertheless, the planning period should be carefully selected in consideration of several following factors:

- Most sewers have a service life of over 50 years;
- Concrete tanks and structures have a service life of 30 to 50 years, or more, depending on corrosion potential and appropriate design considerations;

- Field instrumentation and control devices (e.g., flow monitors, SCADA system equipment) have a service life of 5 to 20 years;
- Mechanical equipment (pumps and blowers) have a service life of 10 to 20 years;
- Utilities develop 3- to 10-year capital improvement plans;
- NPDES permits are typically renewed every 5 years;
- Population and growth projections are "checked" by census very 10 years;
- Many budget decision authorities (elected officials) have a 2- to 4-year election cycle;
- Capital project implementation can take 2 to 10 years from plan to startup; and
- Many large or complex projects, including some long-term control plans for CSOs, may take more than 10 years from plan to startup considering the policy, political, permitting, and financial issues that must be addressed during implementation.

Ultimate or build-out flows and loads conveyed to the WRRF from the designated service area should be considered to provide that the WRRF site can accommodate expected future flows and loads before spending considerable sums for construction.

Many utilities have opted to produce long-range plans that look at 20-year horizons, with consideration for ultimate or build-out development plans. The plans often contain provisions to review and update the assumptions periodically. Other communities develop plans for multiple scenarios, each reflecting different amounts of growth and, in some cases, reflecting different levels of future wet weather flow control.

The implementation period for improvements will depend on regulatory requirements, system performance objectives, and funding availability.

Updating long-term facilities plans should be a priority for utilities assessing wet weather system performance.

1.2.1.2 Basis for Calculating Future Flows

Changes in dry weather flows resulting from growth (new connections) or changes in land use and densification patterns (possibly increasing or decreasing flows) are addressed in traditional wastewater planning manuals and in textbooks; however, changes in water use patterns, water usage rates from conservation efforts, and an increased focus on recycling and reuse in industrial/commercial facilities may require reevaluation of these assumptions in many communities. Many sewers built in the 20th century were designed with an underlying assumption that most flow rates, therefore sewer size, would be dominated by sanitary or industrial wastewater. More recently, many utilities have recognized that peak flow rates are dominated by RDII, and several have recognized that dry weather flow contains large proportions of dry weather infiltration. Figure 3.4 illustrates the many pathways by which RDII enters a separate sanitary sewer system, and Figure 3.5 illustrates the effect of RDII on peak flows in the system.

In Figure 3.5, almost immediately after the rainfall begins, the total wastewater flow (wet flow) starts to increase because of the added extraneous RDII flow to the dry-weather flow component. In this case, the RDII contribution far exceeds the dry-weather wastewater flow rates, thus having a considerable influence on system capacity requirements.

Several communities have recognized that growth will increase the volume of wet weather overflows by adding base sanitary flow that uses capacity previously used to convey the rainwater–wastewater mixture. Many regulatory-driven programs, for example, have recognized and addressed this fact by including a credit procedure that allows growth only if the increase in base sanitary flow is offset by a more than equivalent decrease in wet weather flow.

Because of the construction grants requirements to eliminate "excess" infiltration and inflow, many utilities have adopted tighter construction standards for new sewers and designed interceptors assuming that the RDII would be controlled. Despite this, the literature and practice cite few cases in which the

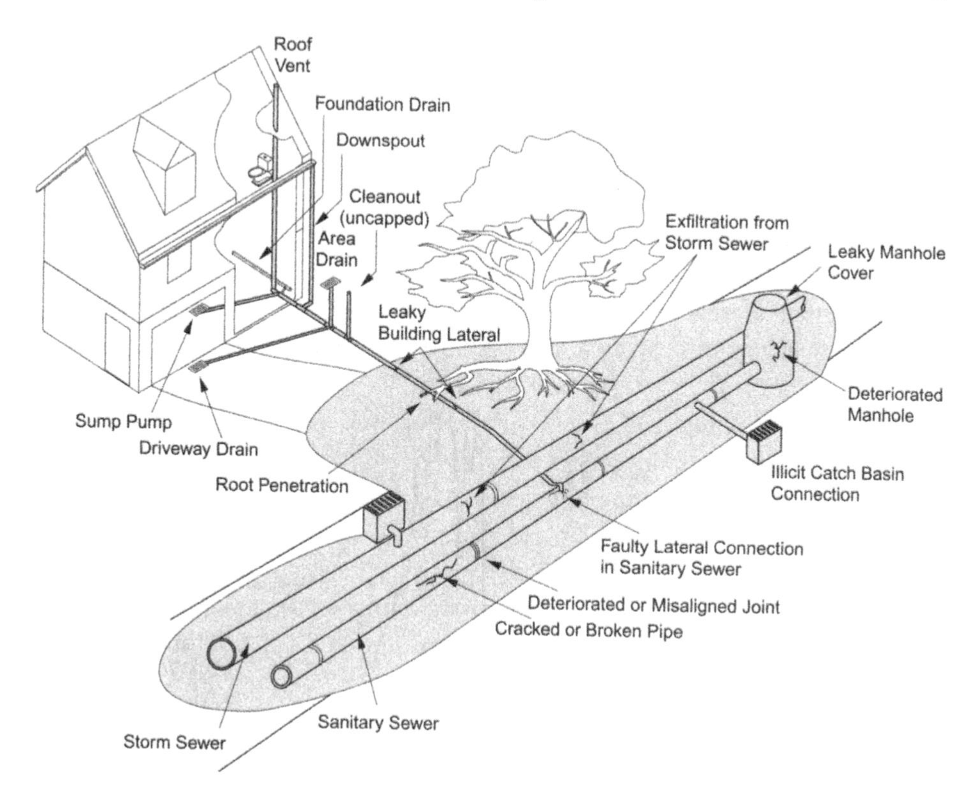

FIGURE 3.4 Pathways for rainfall-derived infiltration and inflow entry to sewers.

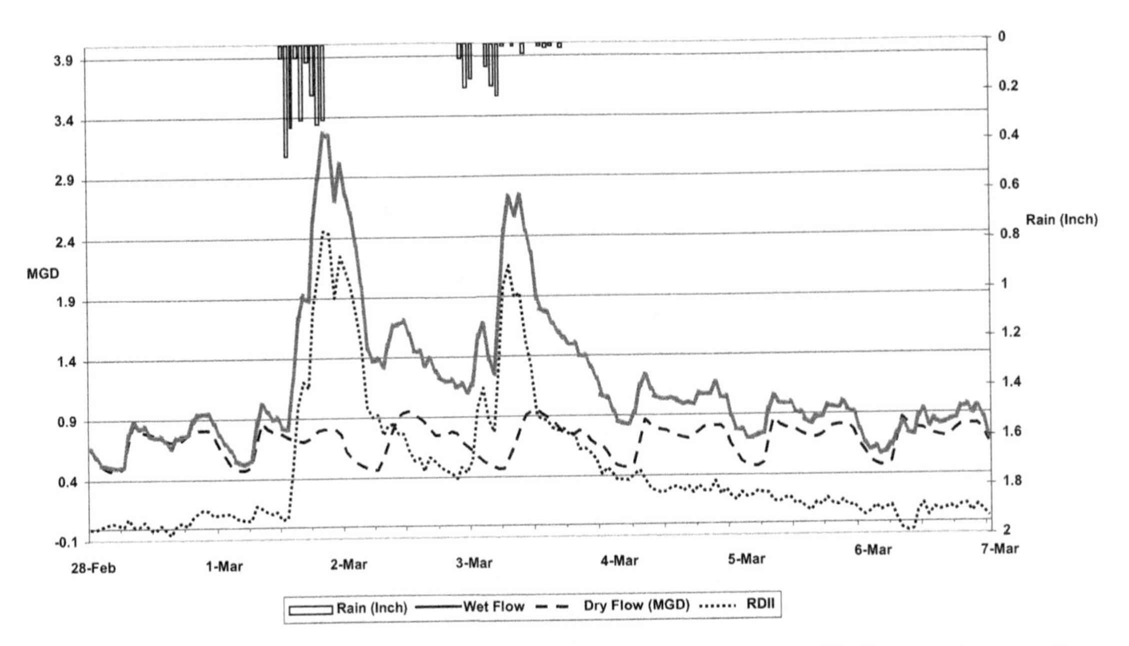

FIGURE 3.5 Contribution of rainfall-derived infiltration and inflow (RDII) to peak sewer flow (in. × 25.4 = mm; mgd × 3785 = m^3/d).

projected decrease in RDII has shown sustained effect at the interceptor level. There are exceptions, however, in which comprehensive RDII removal efforts have resulted in substantial and sustained RDII reduction.

To honor the principle that decisions should be based on fact rather than assumption, one should project future RDII expectations in a manner that is consistent with past experience and the utility's future commitment to infrastructure rehabilitation and replacement. Unless site-specific data exist to the contrary, one can expect that new sewers built today will develop RDII rates similar to the rates apparent in sewers built with similar construction practices and materials from previous years.

Many utilities project RDII on a per capita basis, similar to how they project sanitary flows. Some utilities project RDII flow rates on an inch-diameter-mile basis. Several popular simulation models project RDII on an area proportional basis. Researchers have found that RDII correlates most consistently with the number of manholes (Bennett et al., 1999). When studying existing sewers, the literature emphasizes that RDII is very site specific and variable. That being the case, any projection of future RDII will necessarily make simplifying assumptions that will prove imprecise.

Managing RDII is extremely complex and a marked, site-specific variation in RDII reduction results is often experienced. Consequently, projection of future wet weather flows must rely on realistic projection of the effectiveness of measures throughout the tributary service area that contribute to, or control, whether the rainfall enters the conveyance system. Measures whose implementation and effectiveness should be considered in calculating future wet weather flows include the following:

- Effectiveness of stormwater drainage facilities to redirect rainwater away from sewers, including the effects of stormwater best management practices, such as the use of green infrastructure;
- Building codes to exclude rainwater from sanitary sewers;
- Building code enforcement effectiveness;
- Sewer construction codes to exclude RDII from sanitary sewers;
- Sewer construction code enforcement effectiveness;
- Private property sewer lateral maintenance to avoid RDII;
- Public sewer maintenance to avoid RDII; and
- Other potential measures, such as stormwater reuse and graywater implementation.

Consideration of the maximum flow that can be delivered to the WRRF in the future will be affected by the factors discussed above, but will also be modified by any planned conveyance system improvements. Duration and magnitude (i.e., volume and peak rates) of wet weather flows at the WRRF should be included in the calculation of future flows. The potential for future effects of climate change on weather and rainfall patterns should be considered as part of long-term decision-making.

1.2.1.3 Wet Weather Planning Flows

The first question typically asked by engineers when estimating flows for CSO *long term control plans (LTCPs)* or other wet weather flow planning and designing is, "What is the level of control?" Experience of most professionals is that even within a specific geographic area there is no one design storm appropriate for cost-effective wet weather control planning. Therefore, increasingly wet weather plans are based on simulations of system performance over multiple storm events or flows occurring over a period of multiple days, months, or years. A storm with a specified return frequency for a short-duration event (e.g., 1- to 3-hour peak intensity) may be appropriate for evaluating conveyance and treatment alternatives, but it may be misleading in evaluating storage effects. Conversely, a storm with a specified return frequency for a long-duration event (e.g., a 24- to 48-hour volume frequency) would be useful for storage analysis but could be inadequate for conveyance evaluations. Either storm may be inaccurate for estimating annual loads and discharge occurrences, or the effect of in-system storage on annual overflow volumes. For this reason, multiple storms or a planning period approach should be used to determine system performance over time.

If a design storm is externally specified (mandated by regulation or past practices), facilities designed to handle that storm may have an undefined likelihood of failure. Conveyance designed for a given frequency design storm may fail more often than implied from the design storm frequency if the design storm distribution does not match peak intensity for that frequency for all durations (e.g., the upper portions of the system may fail during a 5-year, 1-hour storm if designed for a 5-year, 24-hour storm). Other influences that make likelihood of failure difficult to predict include the influences of antecedent moisture conditions, influences of stages and flows from adjacent waterbodies (that may be affected by outside factors), and seasonal influences on rainfall patterns and groundwater elevations. Storage designed based on a single-design storm event will probably fail more frequently than predicted if storage drain times and back-to-back storm performance are not considered. Water resource recovery facilities designed for a specific storm will suffer the similar risk of failure under more realistic conditions of multiple storm events under varying moisture and seasonal conditions. Furthermore, if any one of the components fails during a given storm, other portions of the system may fail even if appropriately designed for that storm. For example, if a storage basin is filled from earlier storm events and does not have sufficient time to drain, then the conveyance system may not be able to be relieved to storage and thus could fail due to surcharge.

Several investigations have evaluated the relative benefits of design storm versus continuous hydrologic approaches for evaluating stormwater in urban areas. Most have concluded that continuous methods are technically more appropriate. Continuous simulation methods may be more time consuming, but the time spent in using the methods is compensated by the time saved in justifying assumptions regarding the dependence or independence of events or the costs saved in directly analyzing the effect of wet weather controls on the more common small events. The continuous simulation approach better reflects the effects of seasonal groundwater influence, antecedent moisture conditions, and other outside influences such as river stages and flows of adjacent waterbodies that can be controlled based on water supply and flood control requirements. Use of continuous methods for estimating wet weather event frequency, volume, and loads is recommended.

The use of selected storms from the long-term record for detailed evaluations of sewer relief needs is also good practice for analysis of system components in which a specific peak flow capacity must be accommodated, such as a pumping station. In cases in which model simplification may be required for continuous simulation run times to be reasonable, a single-event simulation is recommended to confirm detailed system hydraulics. Therefore, both single-event and continuous simulations are sometimes appropriate for wet weather planning of system improvements.

One demonstrated approach uses a continuous-precipitation time series that is representative of the range and variability of precipitation patterns experienced in the study area. Long-term (±40-year) hourly or 15-minute precipitation records are available from the National Oceanographic and Atmospheric Administration for first-order meteorological data stations.

These records can be used in their entirety, or a statistically representative shorter time series can be extracted from the long-term record. Many studies identify a shorter period (e.g., 3 to 5 years) demonstrated to be statistically similar to the long-term record in terms of the following:

- Daily rainfall volume frequency,
- Storm duration distribution,
- Storm volume distribution,
- Storm maximum intensity distribution,
- Storm average intensity distribution,
- Daily volume distribution, and
- Annual volume distribution.

To account for snow and snowmelt potential, the long-term precipitation trace can be transformed into an equivalent moisture trace using hydrologic algorithms for snow accumulation and snowmelt.

The best approaches will consider system performance under different scenarios, including both single-event and continuous simulation scenarios. Although some states and U.S. EPA regions have chosen a specific target storm, there is no evidence that this results in optimum system performance or the most effective solutions. Managing wet weather flows is challenging and the search for solutions should recognize variability and system performance under different conditions as critical to the selection of the best alternatives.

1.2.2 *Establish Performance Objectives*

Chapter 2, Section 2.0 advocates a risk-based approach to confirm initial performance targets for management of wet weather effects. It provides guidance for applying risk management techniques, in cooperation with wastewater system stakeholders, to identify mutually acceptable performance standards. This particular protocol section briefly discusses how the risk-based approach applies to the following:

- Defining a long-term design capacity basis,
- Regulatory requirements,
- Bacteria or pathogen reductions,
- Capacity component allowance or techniques,

- Maintaining design capacity, and
- Management activities.

1.2.2.1 *Basis for Long-Term Design Capacity*

Proper wastewater system capacity performance objectives and targets should reduce the risk of negatively affecting public health and the environment. Future wet weather effects and the risks they present must be addressed in the development of wastewater system capacity. The capacity performance objective should be attained as soon as possible (early action projects in facility planning) and maintained consistently as the conveyance system evolves. The initial level of protection against risks may be lower than the ultimate goal, but planning should address long-term progress and measurement of progress toward that goal.

The design capacity basis of the WRRF often is defined by the NPDES permit conditions. Some permits include flow limits as well as concentration-based or mass-based standards for pollutants. WRRFs are designed to meet the permit conditions at the projected flows and pollutant loads over the life of the planning period.

1.2.2.2 *Regulatory Requirements*

The CWA and derivative regulations are established to protect the health of the public and of the environment. Stakeholder values and corresponding objectives should be defined using the risk-based approach to support the primary objectives of the regulations. The performance risk-based approach provides an objective means of monitoring progress and direction toward the CWA goals of mitigating wastewater system overflows and meeting water quality standards. U.S. EPA never implemented a draft policy proposed in 2005 for managing peak wet weather flows at WRRFs servicing separate sanitary sewer collection systems, which would have required a "utility analysis" with the objective of minimizing diversions around secondary WRRFs and maximizing treatment of wet weather flows (U.S. EPA, 2005b). Nor has U.S. EPA developed any other guidance on evaluating wet weather management alternatives. In the absence of this guidance, WRRFs must continue to work on wet weather flow management. The performance risk-based approach was consistent with the goals of the proposed utility analysis and remains a valid tool for developing and analyzing proposed alternatives.

1.2.2.3 *Bacteria or Pathogen Reductions*

Bacteria and pathogens are ubiquitous; hence, elimination of risk of bacteria and pathogens is not possible. The risk-based approach encourages identification and control of the sources of bacteria and pathogens that result in the greatest risk. Thus, the risk-based approach focuses attention on measures that most efficiently reduce the risks from bacteria or pathogens, rather than requiring expenditures to control sources that may result in minimal risk. Communities in which streams exceed bacteria water quality standards (WQS) during dry weather may prioritize reducing the risk associated with failing septic tanks, whereas communities in which streams exceed bacteria standards only after rainfall may prioritize wet weather overflow reduction.

Fecal coliform, *Escherichia coli*, and enterococci are often cited in NPDES discharge permits as indicator organisms of fecal contamination. These indicator organisms are typically the standards by which the pathogen removal effectiveness of a WRRF is measured. Although there is much research and discussion of measuring and monitoring for pathogens reflecting more complex indicator organisms or specific organisms of concern, the established standards are reflected in almost all NPDES permits.

1.2.2.4 Capacity Component Allowances or Techniques

One approach to wet weather capacity planning is to include an allowance for wet weather flows in the design capacity for each component of the wastewater system. This is, indeed, the approach advocated by the widely used Ten States Standards. The risk-based approach provides techniques that consistently define the magnitude of such allowances. The allowances may vary by component class, dependent on the severity of the risk associated with the component. Therefore, the higher the risk associated with failure of a component, the higher the safety factor or level of control that should be used in sizing the component. For example, if the risk from untreated wastewater is greater than the risk from partially treated wastewater, the allowance may be greater in setting capacity for sanitary sewers than in setting capacity for secondary treatment units.

1.2.2.5 Maintaining Design Capacity

Capacity designed and built into the sewers can be lost to sedimentation, increased roughness associated with corrosion, decreased pumping capacity associated with pump failure or wear, increased RDII associated with decay of the sewers, or many other conditions and consequences that could be minimized by better maintenance. The design capacity is reduced to an effective capacity. The risk-based approach encourages identification, funding, and scheduling of maintenance activities that have the greatest potential to reinstate conditions whereby the effective capacity significantly equals the design capacity to reduce risks to key public values. Thus, although the public values and experiences the aesthetics of public grounds, the risk-based approach might recommend prioritization of sewer cleaning more than prioritization of landscaping because clogged sewers are a risk to higher priority values of public health and safety. Note that in a poorly managed system, the risk-based approach may prioritize capital planning to replace damaged assets, whereas in a proactively managed system, the risk-based approach may prioritize inspection of private property to control RDII sources.

Effectiveness and efficiency of conveyance systems, treatment processes, and equipment can be reduced over time, reducing capacity from design values. It is important to understand the design capacity of each unit process and to regularly check the actual capacity under operating conditions to identify any reduction in capacity and to implement measures to restore that capacity.

1.2.2.6 Management Activities

Wastewater system management is a continual effort to prioritize, assign, and measure the effectiveness of scarce resources (dollars and staff). If wet weather planning uses the risk-based approach to set performance objectives, then it will provide managers with both a set of defined objectives and a set of appropriate measures of achievement of those objectives. These objectives and measures will be aligned with the stakeholder values and objectives, including those of the utility management.

Existing management practices include a variety of internal and external reporting activities. One utility developed an Excursion Tracking System that produces reports of SSOs and other system anomalies in a database that incorporates the reporting requirements of its state regulatory authority. A utility's excursion tracking system should be specific to the utility and local regulatory language.

1.2.3 *Evaluate Alternatives to Optimize Wet Weather Flows*

Various alternatives are available for managing wet weather flows in WRRFs. Control alternatives may be classified in general categories such as source control, system modification, control and management, storage, and treatment. Maximum benefits are often realized when addressing wet weather flow issues via a comprehensive approach that views conveyance and treatment systems within the same framework. This maximizes the number of opportunities available for controlling and treating wet weather flows, consistent with an integrated wastewater system approach.

1.2.3.1 Conveyance

The literature describes numerous alternatives for improving the conveyance system performance during wet weather. For combined systems, long-term control plan documentation typically provides comprehensive summaries of available technologies, their implementation requirements, and costs.

Over the past several decades, the recommended protocol for identifying and evaluating conveyance alternatives to manage wet weather flows has included numerous practices aimed at infiltration and inflow or combined sewer overflow reduction. Other technical guidance documents provide more specific details on their implementation. The following subprotocols summarize some of the key considerations in major classes of potentially applicable technologies.

1.2.3.1.1 Flow reduction to or in conveyance systems

Stewards and practitioners of conveyance systems should acknowledge the simple fact that stormwater contributes to overflows. Exclusion of stormwater from the conveyance system is the intent of separate sanitary sewer systems. The systems are built with no intent to capture stormwater, although often with an allowance acknowledging that some stormwater will be captured.

Local codes and construction standards include a variety of provisions aimed at excluding stormwater-derived inflows. Training and enforcement of the local codes is essential to minimizing the amount of stormwater that enters the sanitary sewer system. However, the bulk of stormwater entering sanitary sewer systems is from older development in which construction standards were less stringent than today and the integrity of the public and private pipes has degraded over time. As discussed in Chapter 2 of this guide, U.S. EPA released its IPF (2012), which provides guidance on incorporating alternatives and priorities for stormwater controls into a comprehensive wet weather management plan.

In communities that also have combined sewers, builders and inspectors alike may be confused about which drains may, or may not, be connected to a particular sewer. Although the intent of combined sewers is to capture rainwater, several communities now require disconnection of rainwater inlets during redevelopment in the combined sewer areas, if the disconnection will not cause surface flooding or icing problems. Other communities are using sewer separation or source control methods, such as green infrastructure and stream separation, to remove stormwater-derived inflows.

Many communities have tested infiltration and inflow reduction measures in separate sanitary sewer systems. Others have implemented sewer separation or partial separation in combined sewer areas in a similar attempt to remove RDII. The sources of the RDII are often on private property. Several studies have found that RDII reduction efforts have limited success unless public and private source categories are controlled in the effort. If only some of the sources are controlled, then the stormwater often migrates to the nearest uncontrolled source resulting in little, if any, benefit to the wastewater system.

Similarly, many wastewater utilities have observed that inadequate stormwater facilities make it very difficult, if not impossible to exclude the stormwater from the separate sanitary sewer system. Significant areas of the sewer system can be within the boundaries of the 100-year floodplain and subject to submergence under extreme conditions.

Efforts to be considered to exclude stormwater from the separate sanitary sewer system include efforts to improve the effectiveness of those items listed as pertinent to projecting future wet weather flows (Section 1.2.1.2), such as the following:

- Improvement to stormwater drainage facilities to redirect stormwater away from sanitary sewers, including incorporation of stormwater best management practices, such as source control, green infrastructure, or stormwater reuse;

- Building code enhancements to exclude stormwater from sanitary sewers;
- Building code enforcement enhancements to improve effectiveness;
- Sewer construction code enhancements to exclude stormwater from sanitary sewers;
- Private property sewer rehabilitation and maintenance to avoid stormwater I/I;
- Public sewer rehabilitation and maintenance to avoid stormwater I/I; and
- Public education on keeping stormwater separate from the sanitary sewer system.

Stormwater naturally seeks the path of least resistance, and, unless the accessible entrance points to the sanitary sewer system are sealed, stormwater will continue to enter. Therefore, for an RDII reduction effort to be successful, these stormwater-accessible locations to the sanitary system, both on the public and private side, need to be identified and evaluated for removal. However, since the U.S. EPA Construction Grant years in the late 1970s and early 1980s, the industry has realized that removing 100% of the RDII is an impossible goal. Therefore, a utility planning to implement an RDII reduction program in the separate sanitary sewer system should address the following key questions with the following possible approaches:

1. Should additional conveyance and treatment be constructed instead of an RDII reduction program? A business case could be developed comparing the overall wet weather program cost of conveyance and treatment to RDII reduction.

   In this business case analysis, it would be important to consider the baseline cost of ongoing asset management of the existing sanitary sewer system and existing storm sewer system (or lack thereof). Often, the investments needed to reduce RDII are the same investments that would likewise be needed to properly maintain and proactively rehabilitate aging separate sewer system infrastructure. Therefore, investments in new conveyance and treatment may be additional to instead of in lieu of rehabilitation and renewal of existing assets to reduce RDII. As asset management practices are implemented, RDII should be further reduced or stabilized.

   At a minimum, total costs need to be considered on both sides of the comparison via the following equation in which the right side includes the costs of implementing an RDII reduction:

$$\$CT + \$SanAM + \$StormAM = \$PublicRDII + \$PrivateRDII + \$SanAM + \$StormRAM + \$ReducedCT,$$

Where,

$CT = conveyance and treatment cost;
$SanAM = ongoing sanitary sewer asset management cost;
$StormAM = ongoing storm sewer asset management cost;
$PublicRDII = public RDII reduction improvements cost;
$PrivateRDII = private RDII reduction improvements cost;
$StormRAM = storm sewer system improvements to address the removed stormwater and existing asset management cost; and
$ReducedCT = reduced conveyance and treatment cost.

Analyzing the total costs on both sides of this equation will then provide a utility the information needed to inform a business case decision on the cost-effectiveness of RDII reduction, particularly private property-focused RDII programs. Business case evaluations favorable to RDII reduction can then be used to gain endorsement from the local elected officials, the public, and other stakeholders to advance the program.

2. What is the utility's driver for reducing RDII? Is the focus on LOS objectives, such as reducing SSOs, creating capacity to accommodate growth, reducing flows to a WRRF, or other drivers? After the utility has identified its LOS drivers, it can set the appropriate RDII performance reduction measures and targets. For, example, if reducing SSOs is the objective, understanding how much stormwater needs to be removed to eliminate local SSOs will help the utility select the appropriate RDII reduction performance measure.
3. How will private property RDII reduction be funded? The business case development described in Question 1 can provide useful information for a utility to use to evaluate against LOS objectives and whether public dollars should be spent on private property to pay for all or a portion of the private property cost.
4. How will private property fixes be implemented? A business case may reveal it is more cost-effective to remove the public and private RDII than to convey and treat it. However, consideration needs to be given on how a utility will successfully remove RDII from private property. The following are additional questions to consider:
   a. What is the primary source of private property RDII? Is the flow coming primarily from degraded private sewer laterals or direct inflow connections, such as downspouts, area drains, sump pumps, or foundation drains?
   b. What types of ordinance requirements and enforcement are needed to compel property owners to complete their fixes? Alternatively,

what types of incentive programs should be provided by the utility to encourage property owners to voluntarily complete their fixes?

c. What level of interaction is needed between the public and the utility to facilitate the private property fixes? Should the utility be involved in helping the private property owner design and/or implement the proper fix? Is the utility willing to perform the level of public interaction needed?

d. Can the fixes be made to remove the RDII without causing unintended consequences, such as flooding, ponding, or wet basements? Does the utility need to install a new storm sewer system to properly remove some of the RDII connections, such as driveway or area drains?

e. What types of contractor programs will be made available to the public? Will the utility preapprove contractors? Will the utility perform the work on private property through its own contract or require the property owners to hire their own or a preapproved contractor?

f. What type of construction inspection program is needed by the utility to make sure the fixes have been properly performed?

g. What type of follow-up program is needed to provide that fixes stay in place and don't cause unintended consequences, such as flooding or surface ponding (i.e., lateral condition is maintained and disconnected inflow sources are not later reconnected)?

5. If new storm sewers are needed to disconnect private property connections, how will the utility handle this construction?
6. Is there an opportunity to perform integrated planning to address both the storm and sanitary sewer obligations of the utility as part of the RDII reduction projects? Can green infrastructure be incorporated into the solution to address stormwater water quality and peak flow effects?
7. How will the utility adequately and frequently keep the property owners and other stakeholders informed and engaged in the program?

1.2.3.1.2 Retaining flow in system

Often, the upstream portions of the wastewater system have adequate capacity to convey peak wet weather flows that exceed the capacity of the downstream components. When this is the case, the cost-effective solution may be to retain some flow in the upstream systems and release it later. Systems naturally retain some flow in that the flows get deeper and the velocities slow as the outlet sewers surcharge. This flow attenuation can be enhanced by oversizing upstream sewers or by incorporating mechanical devices (vortex flow control devices, orifices or gates) to force the flow retention. Inclusion of automated controls on the devices allows for real-time control (RTC) that can ensure that available storage in the upstream sewers is used to protect other portions of the system. Fully dynamic hydraulic models are used to evaluate the opportunities for flow attenuation and RTC.

1.2.3.1.3 Tunnels

Tunnels represent a less disruptive means of constructing sewers because they do not require large open cuts for the entire length. Tunneling technology is rapidly evolving to the point where it is now often a competitive alternative, particularly in densely developed areas. Tunnels are particularly competitive for relief sewers where connections need only be established at a limited number of points.

Tunnels are also effective means of constructing storage capacity. Each sewer has some dynamic storage capacity, attenuating the downstream peak flows. The larger the sewer, the greater the downstream peak attenuation. If sewers are large enough, the storage attenuation of downstream peak flows can be significant, and often can be maximized using real-time controls to force additional storage. Tunnels can be used for in-line conveyance and storage of wet weather flows, in a similar manner as conventional sewers provide some attenuation of peak flows. Tunnels for wet weather control can also be used in an off-line configuration to store overflows or excess wet weather flow during storm events and then to dewater the stored flow to a WRRF after the storm events end. Several utilities have made extensive use of tunnels and RTCs to improve collection system performance during wet weather (Schultz et al., 2004). When evaluating tunnels compared to other wet weather technologies, life-cycle operation and maintenance (O&M) costs, including energy consumption, and the need for more complex methods for inspection and cleaning of tunnels should be determined. Site-specific factors, such as surface disruptions at shaft locations, should also be considered.

1.2.3.1.4 Added capacity from new or expanded existing components

The most cost-effective means of adding capacity is often to expand or parallel the existing wastewater system components. Such twinning or expansion has the advantage of maximizing use of the existing utility infrastructure (maintenance right-of-way, electrical connections, etc.). It also is most readily incorporated into the existing operation and maintenance routines.

1.2.3.1.5 Operations- and maintenance-related capacity increases

One of the least-cost means of "increasing" system capacity is often to confirm that the existing facilities function as intended. Common operation-related "capacity thieves" include the following:

- Ineffective backflow devices (flap gates, tide gates);
- Out-of-service pumps;
- Clogged orifices or siphons;

- Debris, roots, grease, or grit accumulation in pipes;
- Broken pipes;
- Missing manhole lids; and
- Deteriorated manholes, particularly near the surface.

Preventative maintenance or asset management programs designed to minimize these O&M limiters can be very cost-effective, but are often overlooked. Most planning to improve wet weather performance concentrates on planning for facility improvements that are budgeted through the utility capital improvement plan. Every effort should be made to identify, fund, and staff O&M asset management activities that are deemed essential to maintaining the full operational capacity of the facilities. These asset management programs are also a critical element in RDII reduction as described in Section 1.2.3.1.1, and they provide the added benefit of decreasing the risk of dry weather SSOs.

1.2.3.1.6 Encouraging regulatory or community acceptance

Both sanitary and combined sewers are intended to convey sanitary waste to treatment. With either type of sewer, if wet weather contributes more flow than the sewers can convey to treatment, then some form of relief is necessary. Any form of relief, whether overflows, increased O&M and rehabilitation, or construction of expanded facilities, presents potential for objections from regulatory and community stakeholders.

Regulatory pressure is often a driver for control of overflows, and thus, the regulators will often have direct input into selection of the proposed remedies through enforcement orders or court settlements. Early discussions with regulators are critical to securing regulatory acceptance of solutions. Dialogue should focus on the advantages and disadvantages of different solutions, including the business case outlined above, as well as consideration of the sustainability and net environmental benefit of alternatives and community preferences. An educational program to provide the regulators with an understanding of the community's collection system and the constraints will reduce that risk. The extent to which a partnership and an open relationship with the U.S. EPA or the state can be developed can result in the willingness to obtain flexibility from all parties.

The federal CSO Control Policy (U.S. EPA, 1994) and associated guidance documents emphasize the importance of involving the public and other stakeholders in planning for the management of overflows. Community buy-in is critical, in particular, to the success of necessary rate increases and private property RDII programs.

It is important for the community to understand the scope and costs of the water pollution problem and that we as a society created this problem and it will take involvement from all of us to successfully solve it. Wet weather water pollution is not a utility problem, but a community problem. After all, it is the community being served by the utility that is paying the bill to solve the problem.

Utilities need to educate the community on the problems and most importantly the public health and water quality benefits that will result from the required cost. If the cost of the program does not provide a commensurate improvement in in-stream water quality, then questions should be asked. Can the program be modified to provide a commensurate in-stream water quality benefit? What will the community gain by this expenditure? These types of questions should be fully explored and answered by a utility in preparation for the community outreach program.

Most communities that have implemented effective programs to control the wet weather effects on their collection system, while initially apprehensive of stakeholder involvement programs, now attribute program success and lasting benefits to the value of open and cooperative communication with a wide variety of community stakeholders.

1.2.4 *Evaluate Procedures to Predict Water Quality Effect of Alternatives*

One common objective of wet weather management is to protect the quality of the waters affected by the discharges from the sewer system. Receiving water quality is a reflection of numerous physical, chemical, and biological factors, some of which are natural.

The potential for wet weather discharges from the WRRF to affect the quality of the receiving water is a complex interaction of many factors. Nevertheless, the larger the receiving water (e.g., river as opposed to small creek), the lesser the effect. Simplified calculations can determine which volumes of sewer overflows are likely to significantly affect the receiving streams (e.g., U.S. EPA, 2004).

Protocols to identify the extent to which wet weather discharges from the WRRF affect receiving water quality necessarily address the net effect on the complex receiving waters. Isolating the net effect involves the following three subprotocols:

- Sampling and Modeling,
- Bioassay, and
- Community Input.

1.2.4.1 Sampling and Modeling

Traditionally, receiving water quality standards and related discharge limits were calculated assuming the receiving waters experience constant, steady conditions of drought (dry weather) flow. Historically, most existing receiving water quality sampling programs were designed under these assumptions. Consequently, most existing receiving water quality sampling data are inadequate to reflect the effect of wet weather discharges.

Assessing the receiving water quality effect of existing wet weather discharges requires wet weather synoptic sampling of both the receiving water and wet weather discharges to understand the effects on the receiving water. The results of such sampling can be extrapolated in space (not all locations can be sampled) and time (future conditions) only using simulation models.

Because traditional water quality approaches focused on drought conditions, adapting those traditional sampling and modeling techniques to wet weather requires research and innovation. Many assumptions that work under drought conditions (i.e., the assumption of steady or gradually varied flows) are violated under wet weather conditions. Models calibrated to respond to dry weather discharges are often inadequate to respond to the multiple discharges that may affect a stream during wet weather. The size of the stream and drainage area must also be taken into consideration. For example, when dealing with large streams with drainage areas that cover multiple states, an understanding needs to be obtained of the effect of local wet weather on the waterbody. Wet weather on a local basis may have little to no effect on a large river.

The analyst is cautioned to critically review all assumptions in water quality modeling against the very real variability reflective of wet weather conditions. Use the experience of the Federal Highway Administration and U.S. EPA's National Urban Runoff Pollution studies of the 1980s in identifying statistically valid means of sampling and evaluating the variability inherent in wet weather sampling.

For wet weather water quality modeling, care must be taken to avoid making steady-state assumptions simply because, "that's the way we've always done it". If the assumption must be made, the analyst must test the sensitivity of the results to a range of values and provide the decision-makers a range of possible model predictions and the "most probable" outcome. However, it should be remembered that wet weather sampling and modeling can provide the data or increase the integrity of the extrapolated data to serve as the basis for stakeholder decisions and regulatory negotiations that would otherwise be based on dry weather conditions.

1.2.4.2 Bioassay

Bioassay is an accepted technique for determining the relative toxicity of a given discharge. The technique's applicability to wet weather discharges, however, is questionable. Even short-term tests for acute toxicity presume an exposure of several days' duration, whereas urban wet weather discharges persist for only a few hours and those of constantly varying concentration.

Some states have attempted to overcome the limitations of the bioassay by comparing receiving water fish inventories or benthic invertebrate populations near discharges to those in unaffected areas. This approach has merit but may be limited to those cases in which the only substantial difference between two reference sites is the presence (or absence) of the defined wet weather discharge.

1.2.4.3 Community Input

Community input is essential not only in stakeholder value determination, but also in maximizing the use of existing information. Receiving waters are observed and often monitored by many people who are not associated with the water quality profession, but are qualified to recognize activities that affect receiving water quality. The citizen volunteer who observes that, "those nasty kids dump all kinds of trash in that stream" may also identify the source of an otherwise unexplained source of toxins or bacteria. Some states recognize and use the results of volunteer monitoring by activists groups.

Open and respectful communication with all stakeholders, particularly nonprofessionals, will maximize the use of available information while gaining credibility for the decision process and, ultimately, fostering endorsement of the wet weather recommendations

Substantial community involvement is required by regulations and guidance governing combined sewer overflow long-term control planning. Techniques for maximizing the value of community involvement in wet weather planning have evolved in several communities to the point at which community involvement is a significant asset to the programs. Milwaukee determined that public involvement required equal footing with technical evaluations to further its wet weather management plans (Foy and Sands, 2004).

1.2.4.4 Treatment

Numerous alternatives are available to designers and operators for improving WRRF performance during wet weather. The following subprotocols summarize some of the key features and implementation considerations of potentially available technologies. Each should be evaluated to determine if it is applicable to the specific WRRF and how it would interface with the existing WRRF unit processes and the compliance approach used in relation to the regulatory agencies and the local community. More detailed information about these and other similarly relevant technologies for wet weather applications, can be found in the publications *Best Practices for the Treatment of Wet Weather Wastewater Flows* (WERF, 2002), and *Design of Municipal Wastewater Treatment Plants* (WEF et al., 2010). As part of a holistic approach to managing wet weather flows, consideration should first be given to reducing the intensity of the peak flow and reducing the amount of wet weather flow entering the wastewater system. Volume reduction is only applicable to preventing stormwater from entering the collection system and should be evaluated and implemented as part of a comprehensive capacity management, operations and maintenance (CMOM) program. Storage is an alternative for reducing the magnitude of peak flow rates and first-flush loadings and spreading out elevated flows and loads over a longer period of time. It can be used in the conveyance system and at the WRRF; however, storage does not reduce the overall volume of wet weather flow that needs to be treated.

Although not a regulatory requirement in some states, U.S. EPA's approach to evaluating system capacity has long included adopting a CMOM-type program to optimize the system capacity, management approach, and O&M procedures. This approach can be useful to ensure that unit processes and equipment are kept in the best condition for treating wet weather flows. In addition, using this approach will confirm to the utility leadership, regulators, and the stakeholders that a comprehensive investigation of alternatives has been completed and that the plan for managing wet weather flows at the WRRF is sound.

1.2.4.4.1 On-site storage

If temporary storage of excess wet weather flows cannot be accomplished in the conveyance system, storage at the WRRF may be appropriate. Temporary storage (flow equalization) of excess wet weather flows in basins at the WRRF site, and metering the return flow into the WRRF as the wet weather flows subside, can be a cost-effective alternative in reducing and potentially even eliminating the need for constructing additional WRRFs for infrequent events. Many of the states and regulatory agencies have guidelines for sizing, designing, and operating flow equalization facilities. These guidelines should be consulted for detailed information.

Storage capacity should be evaluated to determine the volume needed, which will be a function of the intensity and duration of wet weather events, as well as of the typical dry weather diurnal flow variations. Determining the total amount of flow to be treated at the WRRF over a design wet weather event period and subtracting the treatment capacity of the WRRF results in the volume needed for equalization. Consideration of the peak flow rate into the basins will help size influent facilities, and desired rate for withdrawal of the flow to be returned to the WRRF will help size effluent facilities from the basins.

Flow can enter the equalization basins via gravity or by pumping, and the same options are applicable for returning flow to the WRRF after the wet weather event has subsided and/or treatment capacity is available. Providing screening and grit removal of the wastewater before entering the equalization basins will result in fewer maintenance problems with the basins.

The equalization basins should include facilities for removal of accumulated solids, mixing of basin contents, aeration of wastewater (if needed), odor control (if needed), and flushing of the basins after the wet weather event. Safe access to the interior of the basin should be provided for maintenance of equipment and removal of large solids not removed by flushing. Sizing of on-site storage facilities are a function of peak wastewater delivery capability from the conveyance system, peak storm event (intensity and duration), and the treatment capacity of the WRRF. It is good practice to divide the equalization basin into cells or compartments that can be filled sequentially to isolate a first-flush load, enhance flexibility, and minimize the time needed to clean up the basin after a small wet weather event.

1.2.4.4.2 Maximizing and protecting existing treatment capacity

After consideration is given to the use of storage to attenuate the effect of wet weather events, efforts should be focused in maximizing the use of existing facilities within the WRRF, while protecting these assets and their longer-term (i.e., beyond the event duration) processing capabilities.

The capability of WRRFs to handle peak wet weather flows can differ greatly depending on actual design and operating capabilities and conditions. Although some general trends can be observed on how much more above average flow an existing facility (or components of it) can handle during wet weather events, significant exceptions exist that must be considered when examining a specific WRRF. The first effort should be to consider hydraulic limitations in conveying these extraneous flows through the facility, identifying restrictions and evaluating potential solutions to address them. This is often referred to as the hydraulic "de-bottlenecking" of a WRRF.

However, in addition to addressing hydraulic limitations, it is important to establish the processing capabilities of existing unit treatment processes individually and of the treatment facility as a whole, and to identify means by which to maximize their ability to effectively and reliably handle temporary wet weather conditions, while protecting the longer-term treatment capabilities of the WRRF. Satisfactory performance at flow rates above the average flow will depend on many factors: the speed and intensity of how the flow rate increases, the temperature of the wastewater, duration of the wet weather event, effectiveness of flow distribution to prevent short-circuiting within unit processes, and others. Options to address limitations could include retrofitting existing systems and making provisions for temporary operational modifications to maximize their processing capacity during these short-term conditions.

For WRRFs with primary clarifiers, a commonly used strategy to maximize the use of existing assets in improving performance and increasing processing capacity is to temporarily convert them to a chemically enhanced operation. Chemically enhanced clarification (CEC) typically increases the throughput of existing primary clarifiers (by allowing operation at higher surface overflow rates) while providing higher suspended solids removal efficiencies. CEC is further discussed in sections below.

From the perspective of protecting long-term operation of a WRRF (in particular those relying on the activated sludge process), the biggest concern for providing biological treatment of wet weather flows is losing biomass in the effluent resulting from high hydraulic and solids overflow rates on the secondary clarifiers. When that happens, there could be an inadequate biomass population returned to the bioreactors and thus will not be available to treat the soluble organics in the incoming wastewater. Because these organisms multiply at a slow rate (days), treatment performance could suffer for days or weeks after the wet weather event has ended. For that reason, activated sludge biomass is treated as a valuable asset, like a structure or pumping system, and is protected by the WRRF operators during high flow events.

Although an activated sludge treatment system (biological reactor and secondary clarifier) does not have an unlimited capacity to receive wet weather flow while separating solids, implementing best practices for design and operation of the system can maximize the use of existing capacity. Providing wet weather flow storage, adopting in-facility flow rerouting (as described below), and temporarily modifying the activated sludge process configuration to reduce solids loading to its clarifiers (by converting to step feed or contact stabilization mode) are all practices used by WRRF operators to protect their biomass inventory. Stress testing is an effective tool to identify capacity constraints so the WRRF can upgrade components and operational practices as needed to increase overall capacity during wet weather events and to protect its long-term treatment capabilities.

1.2.4.4.3 In-facility flow rerouting

Depending on the design of the WRRF, wet weather flow within the WRRF boundaries can be routed around various unit processes if the capacity of those unit processes would be exceeded, if they are considered unable to provide a significant benefit in terms of treatment achieved during these events, or if they could eventually lose the ability to provide treatment on a long-term basis. This practice is sometimes referred to as in-facility diversion. When the effluents of an upstream and downstream unit process are brought together, it is often referred to as blending. This rerouting or diverting of flows within the WRRF is performed to optimize overall treatment in scenarios such as diverting screened and degritted wastewater around primary clarifiers directly to the biological process, or diverting primary effluent around the biological process and combining the two flow streams immediately upstream of disinfection. This operating scenario is shown in Figure 3.6.

Another common in-facility flow rerouting scenario is the process of step-feeding flow to later zones of the bioreactor in an activated sludge process to reduce the mixed liquor concentration and thus the solids loading to secondary clarifiers during wet weather events. Rerouting is also used as a way to protect treatment capacity of biological processes (for example nitrification or nutrient removal capabilities) beyond the duration of the actual wet weather event. The majority of POTWs that participated in a wet weather survey (NACWA, 2003) include some variation of flow rerouting as a necessary part of their wet weather

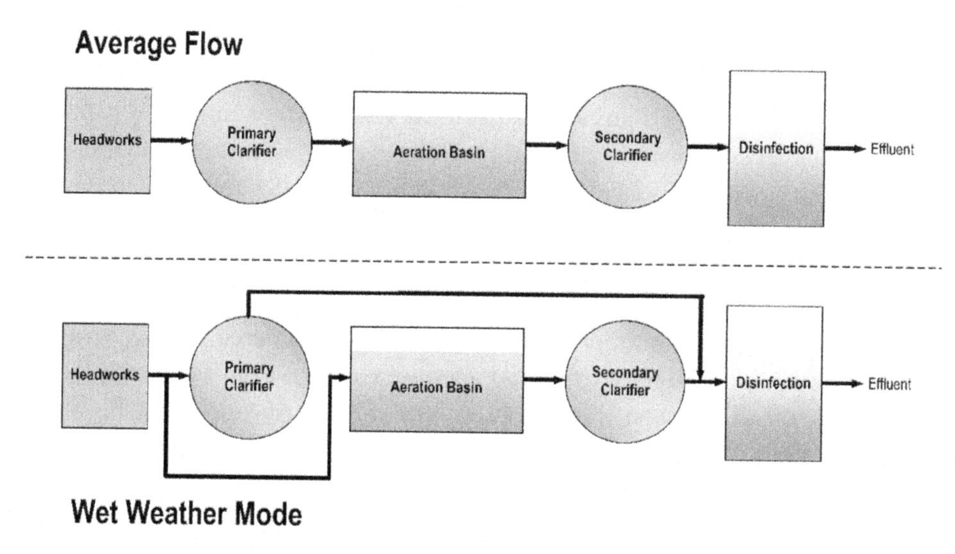

FIGURE 3.6 Split treatment of wet weather flow.

flow management operations. Performance of flow rerouting schemes is site specific and should be verified with on-site testing whenever possible, verifying the capacities of the various unit processes through stress testing.

Flow rerouting/diversion and blending are typically practiced to protect assets from damage, to enhance the overall performance of the WRRF during a wet weather event, and to maintain it after the event. As discussed in Chapter 2, these practices may have implications for compliance with the CWA technology requirements. With the failure to adopt U.S. EPA *Proposed U.S. EPA Policy on Permit Requirements for Peak Wet Weather Discharges from Wastewater Treatment Plants Serving Sanitary Sewer Collection Systems* (2005c), WRRFs are without clear regulatory guidance. These practices should be discussed with the regulators in the context of the WRRF's NPDES permit.

1.2.4.4.4 Providing additional treatment capabilities

In some circumstances, maximizing the potential of existing facilities might not be sufficient to handle wet weather related flows, thus requiring the provision of additional treatment capacity. To a certain extent, some of this additional capacity may be provided by expanding the capacity of existing systems (i.e., providing more of the same); however, planners and designers must recognize that simply increasing treatment infrastructure (e.g., tank sizes and equipment capacity) may not increase the amount of biological treatment that will be provided. Biological treatment may be limited by the amount of biomass that is available with biomass treatment kinetics, which is generally slow in comparison to wet weather flow variations. Capacity can also be provided by means of new installations configured in parallel to existing facilities, requiring influent flow splitting and subsequent blending of effluents. These provisions are sometimes referred to as supplemental or auxiliary treatment. However, one of the most common challenges in providing significant additional wet weather treatment capacity (in particular in circumstances of high peaking factors) is that of availability of space within existing WRRFs. This limitation often dictates the need to rely on compact technologies that operate at higher treatment rates than conventional systems. Compact technologies also offer the advantage of being easier to cover and provide odor control if required.

Some of the more common compact processes considered for wet weather flow treatment applications are fine and micro-screening, vortex/swirl solids separation, chemically enhanced clarification, high-rate clarification, and high-rate filtration. Often, WRRFs handling wet weather flows will also need to provide additional disinfection capabilities. The discussion below presents these technologies in very general terms and discusses their key features and performance capabilities. More detailed information on these and other applicable technologies can be found in the reference documents cited in the beginning of this section.

Fine and micro-screening with openings in the 0.2- to 2.0-mm range can remove significant amounts of suspended solids, achieving removal efficiencies that are close to those achieved by primary clarifiers, but in a fraction of the space. When chemicals (typically a coagulant and a flocculation aid) are added to the influent of these units, the removal efficiencies can be comparable and even exceed those of conventional clarifiers. Providing coarse screening (i.e., >12-mm openings) ahead of micro-screens is still required to remove trash and other large solids to protect these units.

Vortex/swirl-based systems are a technology that has been applied for more than 20 years for wet weather flow treatment. Vortex/swirl separation devices are compact vessels that provide flow regulation and some removal of solids and floatable material. The flow in vortex/swirl devices initially follows a path around the perimeter of the unit; flow is then directed into an inner swirl pattern with a lower velocity than the outer swirl (Figure 3.7). Solids separation is achieved by both centrifugal force and gravity because of the long flow path and inertial separation resulting from the circular flow pattern. The concentrated underflow passes through an outlet in the bottom of the vessel while the treated effluent flows out of the top of the vessel. Vortex/swirl separators for wet weather flow treatment typically would be installed offline and would be empty at the start of a wet weather event. If higher-quality effluent is required, the influent to swirl concentrators can be dosed with a coagulant and a flocculation aid. To achieve the higher level of treatment, two separate stages are required. The first stage allows sufficient contact time to ensure that chemicals are effective (Figure 3.8). With chemical addition, these units have been shown in some pilot studies to achieve removal efficiencies similar to those of conventional primary clarifiers.

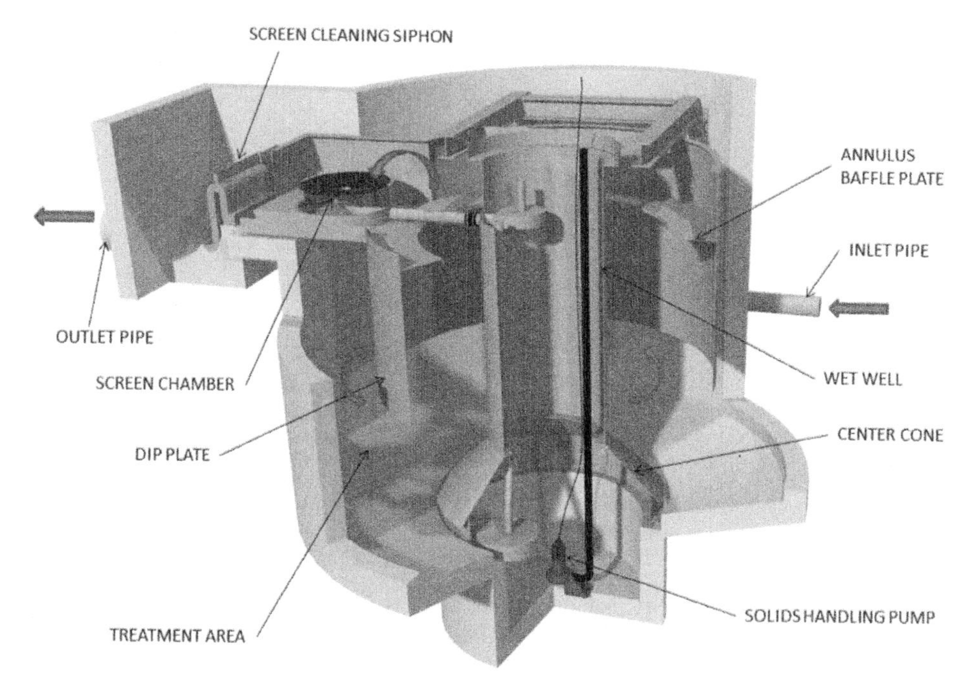

FIGURE 3.7 Diagram of vortex/swirl device (courtesy of Storm King®, H.I.L. Technologies).

As mentioned earlier, a common approach to improving capacity and performance of existing primary clarifiers is to add chemicals. When incorporated in primary clarifiers, CEC is sometimes also referred to as chemically enhanced primary treatment. Initially, work on CEC was done with high dosages of lime at high pH values. Because of the high cost of the chemical and the large amounts of sludge produced with lime, there has been a shift toward the use of iron and aluminum salts as coagulation agents with polymer supplementation to aid in the flocculation process.

The implementation of CEC for primary treatment allows the technology to increase total suspended solids (TSS) removal efficiencies from 55 to 65% (typical for conventional primary clarification) to 75 to 85% as well as increasing 5-day biological oxygen demand (BOD_5) removals. It can also capture and remove heavy metals that could be present in wet weather flows resulting from surface runoff, remove phosphorus, and reduce turbidity. The additional degree of load removal in the primary clarifiers reduces the load to subsequent biological unit processes, potentially increasing the capacity of these unit processes during wet weather events, unless these units are already hydraulically limited as opposed to pollutant loading limited. It also helps to improve the quality of in-facility rerouting of flows if diversion is performed at the WRRF. Conversion of CEC also allows the increase of surface overflow rates, thus significantly increasing the processing

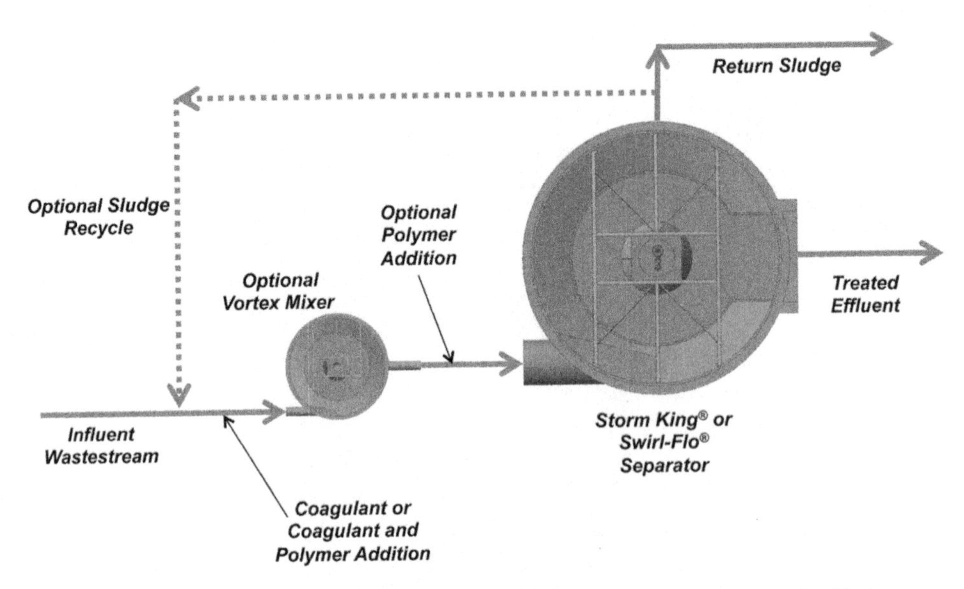

FIGURE 3.8 Vortex separator configuration to increase removal efficiencies (courtesy of Storm King®, H.I.L. Technologies).

capacity of existing primary clarifiers, maximizing the use of existing assets, and in savings in space utilization.

High-rate clarification (HRC) builds on the concept of CEC and is an extremely compact process used to provide high levels of treatment at surface overflow rates 20 to 60 times greater than conventional primary clarification. This is accomplished by adding a coagulant and a flocculation aid to the wastewater, as well as external ballast or solids recirculated from the settling zone to produce dense flocs with high settling velocities. These flocs can be removed efficiently in settling zones equipped with lamella plates or inclined tubes with corresponding high TSS and BOD_5 removal. When the HRC effluent is combined with secondary effluent and disinfected, it can enable the WRRF to meet secondary effluent standards at high peak to average flow rates. As discussed in Chapter 2 and noted above several times, use of a nonbiological sidestream system for excess flows other than those from combined systems may result in a challenge by U.S. EPA as violations of the bypass regulations. A bypass is defined as "The intentional diversion of waste streams from any portion of a treatment facility" (40 CFR 122), and strategies in this guide are intended to help WRRFs successfully address U.S. EPA's challenges.

Figure 3.9 depicts a ballast-based HRC system, and Figure 3.10 presents a sludge recirculation-based HRC system.

Some regulators consider side-stream HRC treatment to be blending, subject to bypass regulations. To avoid this perception, the ballasted-based HRC processes described above have been modified to add a short detention time

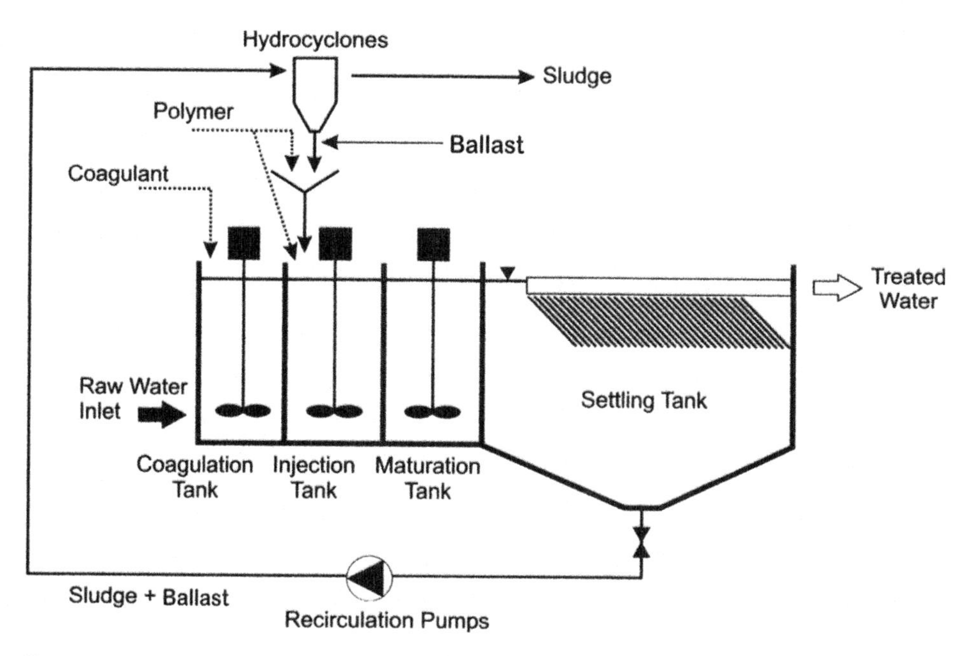

FIGURE 3.9 Ballasted high-rate clarification system (WEF, 2006).

biological reactor to the wet weather flow stream. This brings biomass, typically from the return activated sludge line, into contact with the HRC influent to affect the uptake of soluble organic matter, thus increasing overall BOD_5 removal efficiencies. This adds a biological treatment step to the process, a step that many regulatory agencies prefer for a process to be considered secondary treatment.

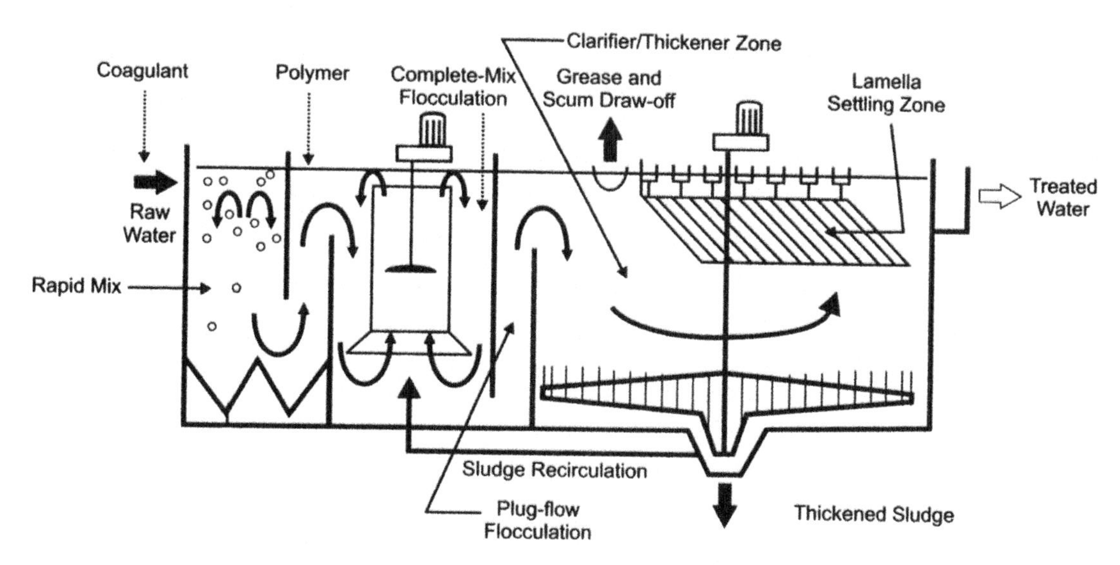

FIGURE 3.10 Sludge recirculation high-rate clarification system (WEF, 2006).

Chemicals are added to provide the coagulation and flocculation of solids before high-rate clarification. Settled sludge is then processed to remove the ballast (which is returned to the process for subsequent reuse), and the remaining material (a combination of chemical, biological, and removed suspended solids from the influent flow) is routed to the activated sludge process of the WRRF for further treatment. This process is sometimes referred to as biologically enhanced high-rate clarification. Figure 3.11 presents a simplified schematic of this process option. Another option is to add magnetite or similar materials as ballast directly into the activated sludge process, increasing the specific gravity of the biomass, and thus allowing the operation of the secondary clarifiers at greater surface overflow rates.

Recent years have also seen the development and more frequent use of high-rate filtration systems for wet weather flow treatment applications. These filters typically rely on other than granular media to operate at higher hydraulic loading rates, which significantly reduces (approximately by 75%) the total surface area requirements as compared to more conventional granular systems. One of these systems is the compressible media filter (CMF). The CMF relies on a synthetic fiber porous material. Because of its low density, the filtration medium is retained between two perforated plates. The filter medium properties, such as effective filtration size, porosity, and depth, can be adjusted in response to changing influent conditions because it is compressible. Key features of a CMF are depicted in Figure 3.12.

Most WRRF effluents require disinfection before discharge to receiving waters to kill or inactivate pathogens that might be harmful to humans. Disinfection

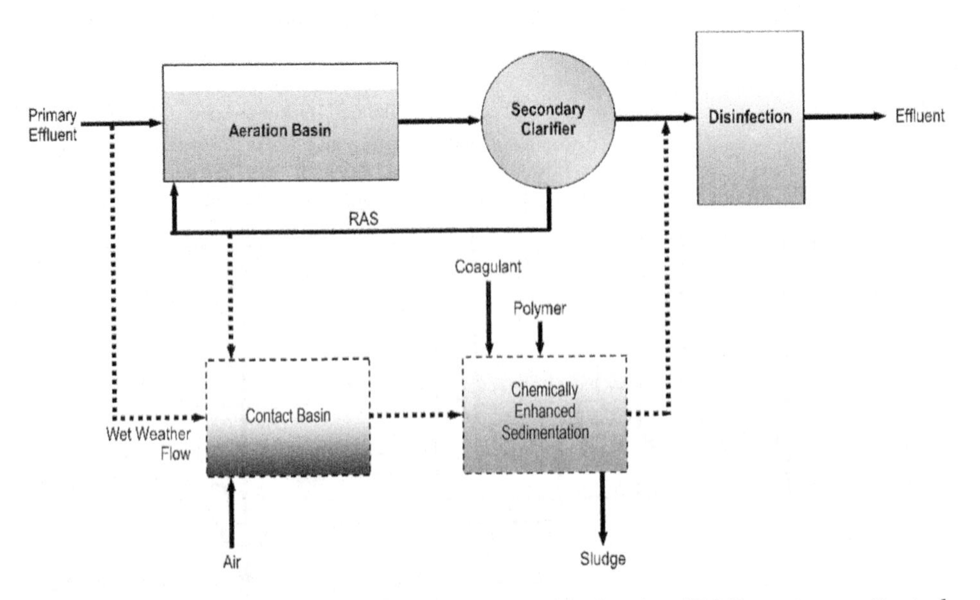

FIGURE 3.11 Biologically enhanced high-rate clarification (RAS = return activated sludge).

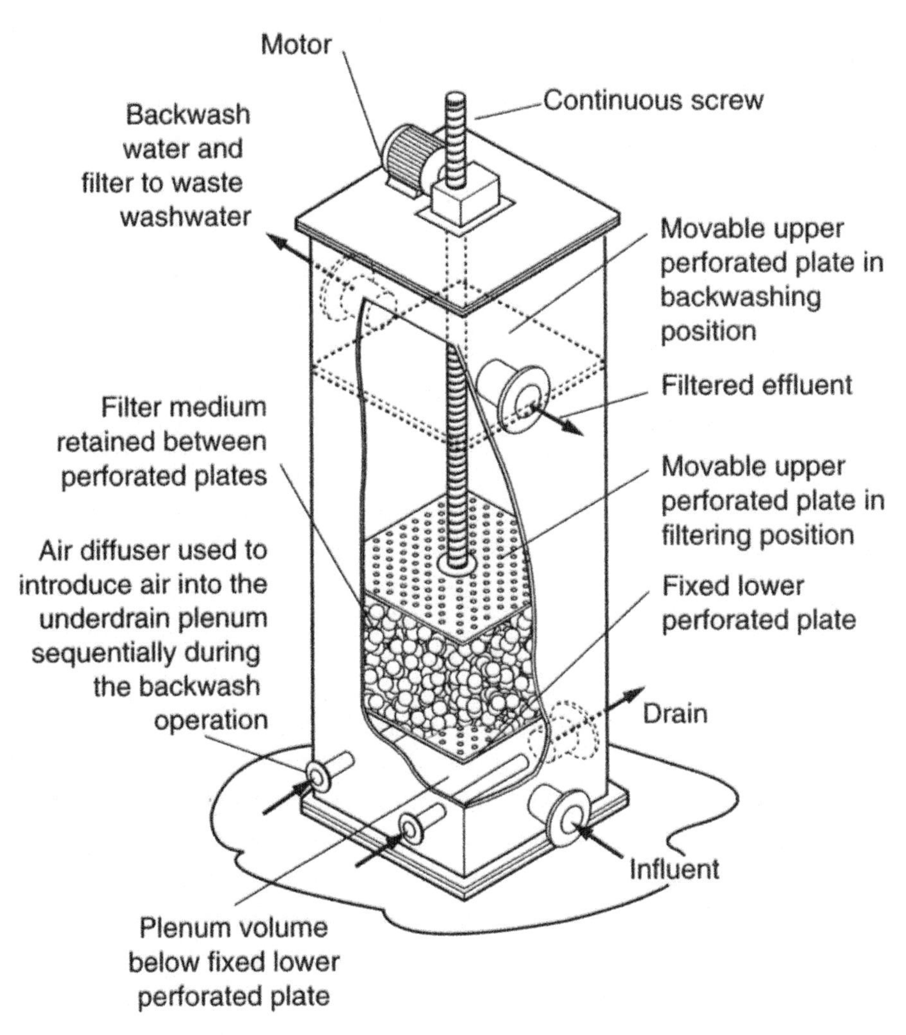

FIGURE 3.12 Schematic view of compressible media filter (WEF, 2010).

standards and design criteria for commonly used technologies, such as chlorine (either as chlorine gas or as liquid sodium hypochlorite solution) and ultraviolet light, are detailed in the literature, and have in many cases standards established by regulatory agencies. In the particular case of chlorine, regulatory standards are established to provide adequate contact of the disinfectant with the wastewater, with the understanding that there are many ways to introduce the chemical and many arrangements available for the contact tanks. Most regulatory design standards require a minimum contact time at peak flow to provide that effluent of potentially poor quality will have sufficient contact with the disinfectant to ensure sufficient kill of the pathogens. In some instances, dechlorination of the effluent before discharge is also required. Full-scale experience and research into high-intensity mixing of the chlorine with the effluent at the point of application has shown that the desired pathogen kill can be accomplished with lower detention times in chlorine contact basins than required by most regulatory standards. This offers promise to provide a higher level of disinfection during wet weather events with more compact solutions and at a lower capital investment at WRRFs. Consideration should also be given to adding chlorine before the flow equalization or clarification stage to provide additional contact time.

It should be noted that although the majority of chemical-based disinfection systems in the United States rely on chlorination for wet weather flow treatment applications, other disinfectants exist, either as methods common in other parts of the world or as emerging technologies. Some of these include chlorine dioxide, peracetic acid, bromine, and ozone. These options are being increasingly considered as potential alternatives to the traditional gas or liquid chlorine-based solutions.

1.3 Select Plan

Numerous alternatives are available for improving WRRF performance during wet weather. The following subprotocols summarize some of the key considerations of potentially available technologies and approaches.

Once a list of technologies with cost and benefit data has been assembled, these technologies should be grouped into a series of alternatives to be evaluated against one another, as previously described in Section 1.2.3. The decision-making process involves evaluating the alternatives using decision criteria that facilitate selecting a targeted performance level as outlined in Chapter 2. Decision criteria that span technical, institutional, social, environmental, and economic areas are developed with stakeholder input and applied systematically. To obtain broad stakeholder approval, the decision process needs to be clear, open, and defensible. The result of this process is the selection of a recommended plan that, if properly implemented, will achieve the targeted performance objectives.

1.3.1 *Select Alternatives*

From the list of technologies previously assembled, determine applicability. Those technologies that are no longer applicable should be documented with the reason(s) why they were removed from the list and are no longer considered. The remaining technologies are then further evaluated to compile and select alternatives.

1.3.1.1 Decision Criteria

The basis for the decision criteria typically includes issues of importance to the permit holder, regulatory agency, and the system users. The decision criteria should address the full triple bottom line (social, environmental, and economic) to enhance the long-term sustainability of the selected plans. Triple-bottom-line evaluation to enhance sustainability is described in many online business references, including Slaper and Hall (2011). It is important to gather data on issues of importance to stakeholders to build support for the ultimate plan. Stakeholders often use a specific decision basis depending on their role in the planning process, be it a ratepayer, owner, regulator, or waterbody user. Decision criteria may include, but are not limited to, attaining water quality goals, public acceptance, effective cost expenditures, reliable operation, regulatory concurrence, and compatibility with existing and future conditions. For example, regulatory agencies are typically most concerned with improving water quality, whereas wet weather stakeholders are typically most concerned with managing water quantity. Quality and quantity management are integral components of every wet weather management plan.

Beyond quality and quantity management, many other issues often prove to be important to stakeholders. These potential decision criteria are often grouped into technical, environmental, and institutional categories. Table 3.1 provides a potential list of decision criteria. The example in Table 3.1 is specific to combined sewer systems. Lists appropriate to particular combined sewer, storm sewer, or sanitary sewer systems would necessarily pare or augment the Table 3.1 list to meet the local concerns.

TABLE 3.1 Typical decision criteria used to evaluate wet weather technologies.

Technical	Environmental	Institutional
Improves quantity management	Improves quality	Acceptable to stakeholders
Maximizes storage (NMC #3)[a]	Improves sensitive areas	Maximizes public benefits
Maximizes conveyance to treatment (NMC #5)[a]	Improves wildlife habitat	Minimizes complaints
Maximizes pathogen reduction	Minimizes environmental effects	Minimizes implementation time
Maximizes existing infrastructure	Minimizes beach closings	Minimizes land requirements
Maximizes reliability	Minimizes fish-harvest closings	Acceptable rate changes; affordable
Minimizes overflows during dry weather (NMC #6)[a]	Minimizes shellfish-harvest closings	
Minimizes overflows during wet weather		
Minimizes solids and floatables (NMC #7)[a]		
Minimizes operational complexity		
Minimizes life-cycle cost		

[a]From U.S. EPA, 1995b; NMC = nine minimum controls.

1.3.1.2 *Tools and Processes*

Once an agreed-upon list of decision criteria relative to the agreed-upon stakeholder values is established, a scoring system must be developed. Scoring systems are either quantitative or qualitative. Quantitative scoring systems are typically more defensible. They are based in numerical scores instead of best judgment. A weighting system may also be developed and applied to the criteria to determine their importance to the overall plan. The Technical Practice Update titled *Wastewater System Capacity Sizing Using a Risk Management Approach* (WEF, 2007) provides an example of listing and weighting criteria for decision support. Optimization techniques can be applied to facilitate the rapid, yet thorough and consistent, evaluation of a large number of alternatives. Optimization tools that have been used in wet weather plan selection range from production theory (MMSD, 2007), genetic algorithms, and other mathematical techniques applicable to the more general multi-attribute utility analysis (Coello Coello et al., 2007).

1.3.1.2.1 Cost curves

A system of cost curves can be developed for technical decision criteria. Figure 3.13 presents a generic cost curve, showing the costs per unit quantity reduced. It is clear from the cost curves where costs begin to dramatically increase for additional quantity reduced—this is what is referred to as the "knee-of-the-curve", beyond which costs continue to increase but with diminishing performance returns.

The CSO Control Policy specifically includes cost/performance considerations using a knee-of-the-curve approach for selecting long-term controls. For instance, the cost/performance consideration may be for the cost of implementing long-term

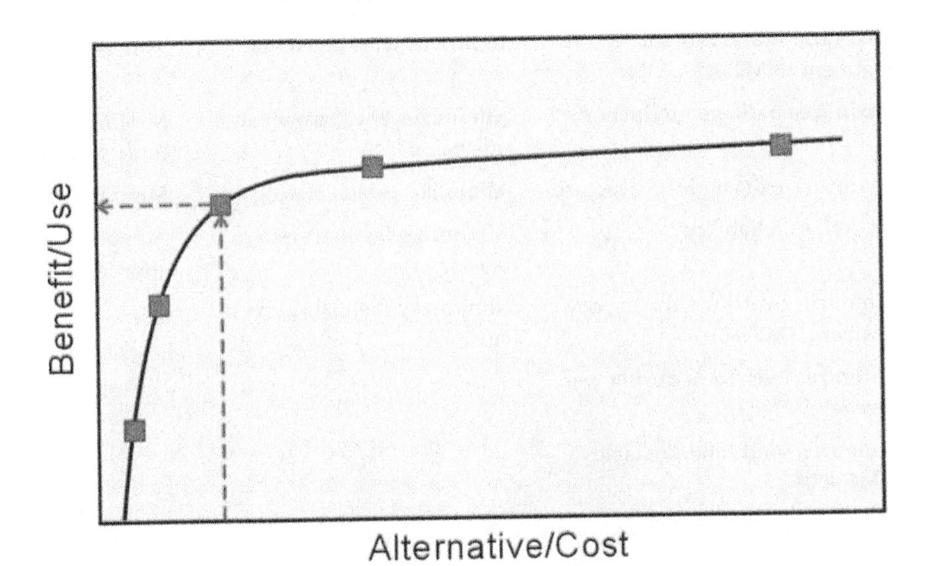

FIGURE 3.13 Generic cost curve.

CSO controls versus receiving water quality improvements to meet a WQS. Figure 3.14 is taken from U.S. EPA's CSO LTCP guidance, which demonstrates how receiving water uses can be used as performance criteria when evaluating technologies. In this case, the uses are those affected by pathogen bacteria discharges, such as shellfishing and beach uses. Benefits are quantified as the decreased number of occurrences of use impairments, rather than quantified reductions of wet weather discharges. This approach to demonstrating benefits may be extremely helpful in a public-participation context, during which stakeholders can realize the projected benefits and their associated costs by readily identifiable values.

The most cost–effective groupings of technologies would be selected and combined into alternatives to meet plan goals. Figure 3.15 is taken from the July 2002 CSO LTCP for the District of Columbia Water and Sewer Authority (DCWASA, 2002). The cost–benefit curve illustrates the levels and costs of controls required for reducing or eliminating CSOs to the Anacostia River. The cost-effectiveness of controls is represented in terms of the percent of CSO volume reduced. Intermediate CSO controls have increasing effectiveness at reducing the number and

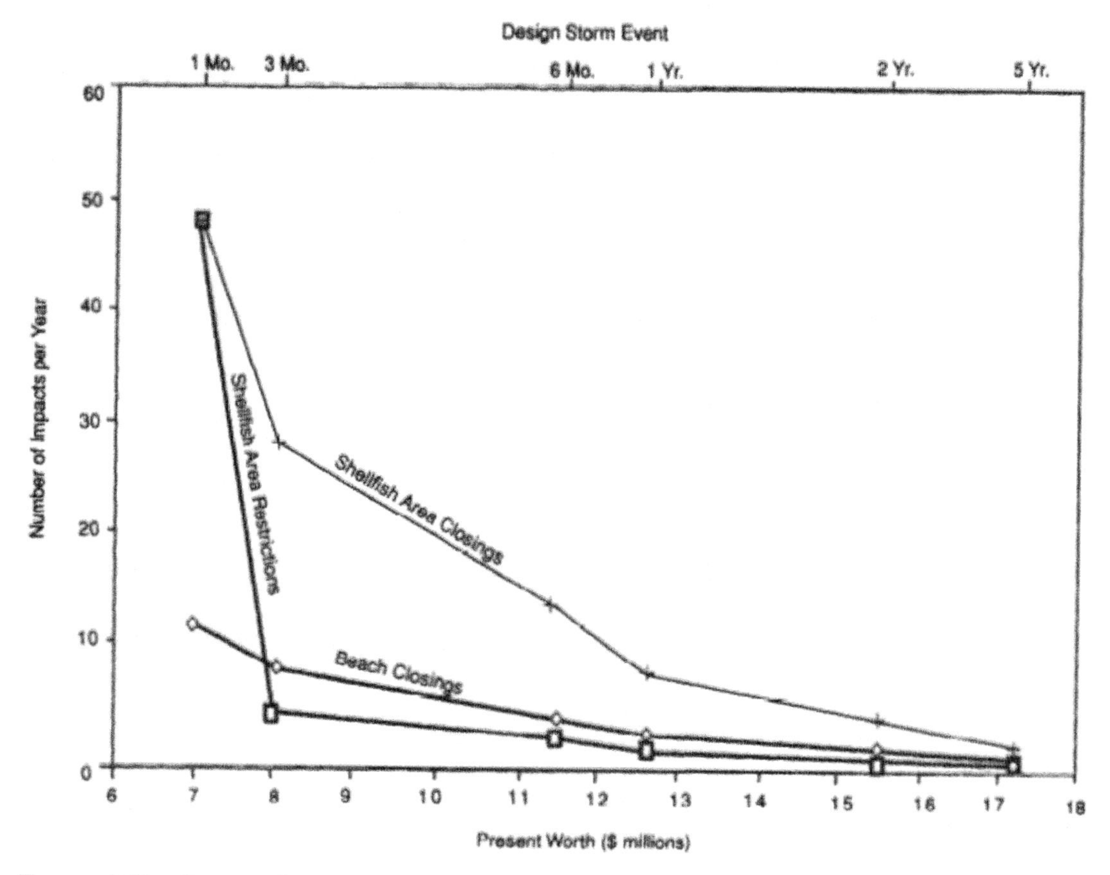

FIGURE 3.14 Cost-performance curves indicating effects on critical uses (U.S. EPA, 1995a).

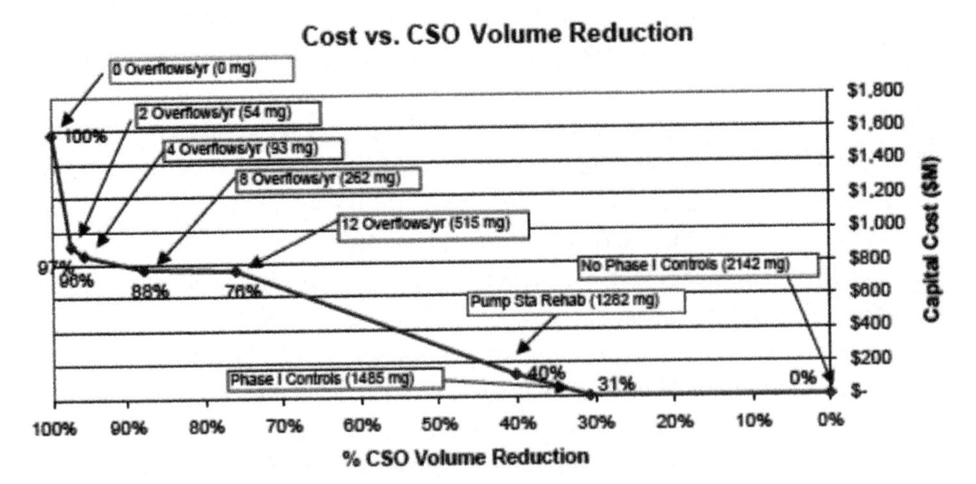

FIGURE 3.15 District of Columbia Water and Sewer Authority's cost–benefit analysis for the Anacostia River (DCWASA, 2002).

volumes of discharges. At the knee-of-the-curve, two overflows per year, the cost curve begins to turn vertical, which implies increasing costs for less benefit.

There is more than one inflection in the curve illustrated in Figure 3.15, indicating several breakpoints in costs/benefits. A cost–benefit analysis may yield cost curves that have several "knees" or none at all. Furthermore, the knee of the curve may not represent the optimum alternative, or combination of alternatives that is satisfactory to fulfill the performance objectives of the POTW, regulators, and other stakeholders. In some cases, all parties may agree on a cost/benefit point beyond a knee that would be satisfactory to fulfill all objectives and ensure acceptance. The knee-of-the-curve analysis is not meant to be an absolute determiner of the solution, but is a tool to be used, with other factors, in making that determination.

1.3.1.2.2 Scales

The non-cost axis on the cost curves need not be inherently numeric, but it does require a scaled range. An independent scoring system offering a scaled range can be developed for any suggested technical decision criterion. Table 3.2 provides example scales for eight criteria used in the Capacity Assurance Project Plan portion of the Metropolitan Sewer District of Greater Cincinnati *Wet Weather Improvement Plan* (MSDGC, 2006).

1.3.1.3 Regulatory or Community Participation

The CSO Control Policy, U.S. EPA's Watershed Approach, and other planning guidelines require or recommend using a public-participation process in developing plans. A public-participation process involves stakeholders, including the affected public, in decision-making for selecting alternatives. There are several

TABLE 3.2 Objective benefit assessment criteria.

Value/Benefit	Factors considered	Benefit level			
		Highest (++)	Moderate (+)	Zero (0)	Loss (−)
Water in basements or streets	Sensitive areas Modeled vs observed	No surcharge of concern under Level A conditions	No surcharge of concern under typical year	2003 conditions	More surcharges of concern than 2003 conditions
Public health risk reduction	Discharges to streams that one could have contact with	No discharges of undisinfected wastewater to or upstream of accessible waters	No discharges of undisinfected wastewater during typical year	2003 conditions	More discharges of untreated wastewater than 2003 conditions
Community standard	Aesthetics —Visible sanitary debris —Odors	No unscreened wastewater discharges above areas where the discharge may stagnate	No unscreened wastewater discharges during typical year	2003 conditions	More debris or odors than 2003 conditions
Environmental benefits	Specific General	Load from collection system overflows less than 5% of total annual or storm-specific load of suspended solids, oxygen demand, or nutrient load to the watershed	Load from collection system overflows pre-2003 conditions of total annual or storm-specific load of suspended solids, oxygen demand, or nutrient load	2003 conditions	Higher risk to aquatic and riparian habitat than 2003 conditions
Infrastructure renewal	Structural Leakiness Current capacity Future capacity	All sewers more than 50 years old subjected to inspection and rehabilitation	Decreased percentage of total collection system more than 50 years old	2003 conditions	Increased percentage of total collection system more than 50 years old

(continued)

TABLE 3.2 Objective benefit assessment criteria (*Continued*).

Value/Benefit	Factors considered	Benefit level: Highest (++)	Moderate (+)	Zero (0)	Loss (−)
Strict interpretation of regulatory compliance	Enforcement Third-party interests	All collection system discharges explicit in permits, no permit violations, no unscreened, undisinfected discharges during Level A conditions	All collection system discharges explicit in permits, no permit violations, no CSOs during typical year, no SSOs during typical year	2003 conditions	More discharge locations not explicit in permit, more frequent discharges at intermittent locations, more permit violations
Flexible with respect to more stringent requirements	Expandable Upgradeable	New collection system sized for highest flows projected by 2022, flexible to increased treatment requirements	New collection system sized for larger flows than expected by 2022, or sized for expansion without substantial cost increases	2003 conditions	New collection system inflexible to expansion or new requirements
MSD* vision achievement	Strategic plan Ratepayer satisfaction	Objectives explicit in the MSD strategic plan fully achieved	Greater achievement of objectives explicit in the MSD strategic plan	2003 conditions	Sewer moratoria issued, ratepayer revolt

*MSD = Metropolitan Sewer District of Greater Cincinnati.

forums in which to present and receive regulatory and community input. Often, a stakeholder group will be formed that includes regulatory and community members who are very informed of the plan in progress and are active participants in its development, progress, and ultimate approval.

Interaction with these groups is performed through either written or verbal communication. Written, possibly electronic, communication documents status and responses and provides time for study and reflection. Written communication can be provided as reports and meeting notes and displayed at the permit holder's place of business, public buildings, and electronically on Web sites. Verbal communication provides a forum for instant discussion and immediate decision-making, occurring typically in a meeting setting. Social media provides an added opportunity for stakeholder communication. Regardless of the media, it is best to document any progress, significant points, and agreements made during the meetings or other communications. Questions and comments must be addressed in follow-up communication.

1.3.2 *Develop Implementation Plan and Schedule*

The implementation plan and schedule is typically developed with the stakeholder group and in accordance first with any legal and regulatory requirements. Community requirements may affect the implementation plan. How the plan is funded will also affect the length of time required to implement the plan. Many plans are developed using a 20-year planning period. Actions are then often grouped into 5-year periods for detailed monitoring as the plan is implemented. Interim milestones can be evaluated and the plan can be adjusted if new technologies are identified, lessons are learned, regulatory changes occur, or other circumstances affect the planning and implementation process.

1.3.2.1 Program Sequencing

A plan consists of several actions that must progress according to a schedule. Actions can be grouped into several categories to determine sequencing, as follows:

- Short implementation time with significant benefit,
- Short implementation time with less significant benefit,
- Long implementation time with significant benefit, and
- Long implementation time with less significant benefit.

Typically, those actions requiring a short implementation time with significant benefit are performed early during implementation to demonstrate immediate return on investment to stakeholders. Actions with longer implementation times with significant benefit have also begun early because they need time to develop further and are critical to meeting plan goals. Those actions with less significant benefit follow. There are often interrelationships between the actions that must be taken into consideration. For example, there may be an action, such as implementation of stormwater detention upstream that greatly reduces flow downstream and for technical reasons must precede another action regardless of implementation time or level of significance. Sometimes the most expensive actions are delayed because actions can result in different conclusions and plans can change, rendering a future obsolete. U.S. EPA's IPF (2012) encourages this type of consideration of the multiple interrelated factors inherent in long-term planning for maximum benefit in achieving community health and water quality benefits.

1.3.2.2 *Regulatory Requirements*

Regulatory requirements are implemented through permit language. Regulations affecting plan implementation can be at the local, state, and federal level and are generally uniform throughout the service area. Typically, the most stringent regulations govern actions.

1.3.2.3 *Additional Legal Requirements*

Enforceable legal agreements (e.g., consent orders), often developed through negotiations, may include specific additional requirements that must be implemented during the planning process. The legal basis of such agreements may also be local, state, or national laws or regulations and may differ from place to place.

1.3.2.4 *Stakeholder Requirements*

Community or special interest group requirements may also affect schedule. For example, dual-use of community land may require wet weather action plan implementation to occur at a specific time to coincide with implementation of another community action; a wildlife habitat may be on the verge of being destroyed if a corrective wet weather action is not immediately taken. These requirements must be known and taken into consideration to properly sequence the implementation plan.

1.3.3 *Secure Funding and Determine Customer Costs*

To implement the plan, funding sources and customer costs must be identified. The amount of available funding from sources other than customers reduces the customer's share. How the plan is funded affects schedule length.

1.3.3.1 *Funding Approaches*

A variety of funding approaches are available: (1) borrow funds that must be reimbursed over time, (2) issue bonds to fund particular projects or programs, (3) be awarded funds for a particular purpose that do not need to be reimbursed, (4) receive funds through customer fees, and (5) develop a new funding program. U.S. EPA provides guidance and resources for identifying funding options in its CSO guidance (U.S. EPA, 1995b), SSO notifications, and other publications.

A common loan program is the State Revolving Fund loan program. Applications are submitted and procedures are in place regarding application requirements and approval. Performance specifications are typically negotiated and detailed before construction begins and then documented within a specified period after construction is complete and the facility has had adequate time to be in operation.

Bonds offer flexibility in financing projects. They are typically structured with long-term retirement schedules and can be used to fund noncapital projects (such as manhole inspection and repair programs) that must be accomplished in a short period of time.

Grants can be provided through a variety of sources. Grants are often very specific regarding the goal of the grant program and therefore eligibility is limited. Grants vary greatly in amounts and availability.

Fees are typically applied uniformly across customer groups, generating a consistent, reliable funding source—smaller amounts of funds generated through numerous customers. Sewer utility fees and stormwater utility fees are examples of fees often grouped in the customer category.

Funding approaches are typically known well before plans are developed; however, throughout the planning process, new funding programs may develop. Historically, customers of sewer systems often were also customers of water supplies and paid one fee for both services calculated based on the metered water used. Wet weather/stormwater programs in some locales are funded by an additional fee developed and assessed based on the amount of impervious cover that contributes proportionally to the amount of runoff that must be managed.

Section 2.2.4 further discusses funding and affordability.

1.3.3.2 Public Involvement

Because the public is affected by both the costs and effects of collection systems and their improvement, it is essential that an informed public be involved in discussions and decisions regarding wet weather strategies. The International Association for Public Participation (www.iap2.org) summarizes evolving resources for public involvement and stakeholder engagement. Public representatives are often part of larger stakeholder groups. At a minimum, public hearings are typically required when public monies are eligible for use. Public hearings include a presentation, followed by a question and answer period. Hearings are often recorded and the minutes become part of the public record and review of the plan. Those asking questions may be given a time limit.

News releases inform the public of plan progress and impending actions and often reach a larger audience because they are not limited to a specific date and time. News releases also may inform the public of the recent discharge untreated wastewater. News releases are word-limited, and therefore concise, and may simply explain where the public can get additional information. New releases can be augmented with Web site updates or other social media broadcasts.

1.3.3.3 *News Media*

The public understanding that is inherent in achieving public acceptance of collection system costs is often affected by news media. News media can provide broad coverage to a wide audience. As entertainment and a competitive industry for viewers, news media focuses on exciting events and generates public interest. The effect of any news coverage is special. It is important to work with the news media to promote benefits of the plan and to outline needs. News media can serve as a valuable independent plan supporter or can damage efforts to gain support if they do not have a good understanding of the plans and proposals. Internet and social media postings are an evolving expansion of available media outlets offering a wider, more rapid distribution.

1.3.3.4 *Special Interest Groups*

Similarly, special interest groups whose goals are aligned with plan goals can serve as an independent plan supporter and source of funds. Special interest groups can generate interest among their constituents and other similarly aligned groups. Likewise, groups can organize to oppose plans based on objections to rates or concern about scope of project or plan elements affecting their neighborhood or other community asset.

2.0 MANAGING

The term *managing* encompasses all utility organizational units and processes. This guide, however, focuses on the *minor processes* of design, construction, and fiscal management. The minor processes are very broad, but as much as possible, the discussion of the various practices is limited to those that relate to managing the issues that arise or evolve around wet weather conditions and activities, particularly wet weather flow issues.

Clearly defined goals enable the development of objectives that result in facility sizing, operational standards, and performance targets that meet the established goals; however, clear planning and implementation goals do not mean management can successfully achieve them. For instance, even with a comprehensive plan and representative goals, management must have proper organizational leadership and funding support to implement the work. Management must be able to anticipate problems, identify the risks, and implement preventive and/or reactive responses to emergencies or system failures linked to wet weather events. Active, adaptive management is essential to efficiently implement the evolving technologies that best meet the changing regulatory and public expectations.

The management practices in this section present what lessons have been learned in the industry on meeting a utility's performance objectives.

2.1 Design

Wet weather design entails the preparation of drawings and specifications that detail the construction and performance requirements for a project. Projects typically are identified in the selected plan (Section 1.3) and comprise judicious combinations of the technologies described in this section. Before starting drawing and specification preparation, preliminary engineering must be completed to develop the basis for the design. Preliminary engineering typically starts with a problem statement that details the purpose for the project. From this statement, the required data are identified, collected, organized, and analyzed. Work completed as part of the preliminary engineering may include the following:

- A review of the existing facilities to include the conveyance system and treatment units to determine their capacity, expected life, operation, and other effects from the new project;
- Establishment of criteria/requirements for correction or mitigation of issues with the existing facilities, I/I;
- Integration of new project with existing facilities;
- Establishment of design standards;
- Alternative analysis;
- Preliminary sizing, layout, and treatment;
- Construction considerations and constraints;
- Cost estimating;
- Environmental effects and assessments;
- Schedule development; and
- Identification of project financial considerations and constraints.

Failure to complete this preliminary engineering not only risks failure of the project but also is detrimental to future projects. The following sections provide an overview of considerations in the design of conveyance and treatment systems.

2.1.1 Conveyance

Design of a conveyance system requires consideration of all components, such as pipes, manholes, channels, and pumping stations along with the influent flow(s) and the discharge flow(s). Influent flows include the sanitary sewer base flow (current and potential future) with inflow and infiltration into the system from groundwater and storm events. Discharge flow(s) include those within the collection system, such as at inline wet weather storage facilities, as well as those that exit the collection system such as at CSO facilities and the WRRF. Accurate quantification of flow(s) (dry weather, average, and peak) requires development of a system hydraulic and hydrologic model and flow and rainfall monitoring, as noted in previous sections. Field surveying to determine invert elevations and pipe slope and mapping the existing collection system may be required along with inspection techniques, such as manhole inspection, smoke/dye testing, and closed-circuit televisions (CCTV) inspections.

This data collection is all part of the system characterization as discussed in Section 1.1 and will provide the basis for the development of design standards for modifications or rehabilitation of existing system or construction of new systems. Data collection and information management recommendations are documented in *Prevention and Control of Sewer System Overflows* (WEF, 2011). Details that should be considered as part of the development of these standards are discussed in the sections below.

2.1.1.1 Construction Materials

Sewer construction materials and specifications are addressed at length in textbooks, manuals of practice, and professional literature. This section merely provides a brief overview for those who lack access to the more detailed references. The conveyance system and its associated component material of construction will vary by application. The following sections describe details to be reviewed as part of the material selection. References to specific pipe materials under one heading do not consider the application of the material under different conditions. For instance, reference to concrete cylinder pipe under the 2.1.1.1.1 Operating Pressure heading does not consider the material's application under corrosive environments.

2.1.1.1.1 Operating pressure

In gravity systems designed for surcharging during wet weather conditions, confirm that the pipe and joints are designed for the head conditions. Prestressed concrete cylinder pipe with a testable joint, ductile iron, or fiberglass pipe may be a better selection for such applications than reinforced concrete pipe or clay pipe. This material selection is also applicable for pipe installed in areas with a high groundwater table where the groundwater pressure may cause joints to leak or fail.

2.1.1.1.2 System life

Specialized coatings, other lining systems, or cathodic protection may be required for manholes, wet wells, and pipes used in some applications to prevent premature system failure. Failure can be a result of corrosion, erosion, superimposed live and dead loads, joint failures resulting from movement or root intrusion, stray electrical currents, or other conditions.

2.1.1.1.3 Flotation

Wet weather events can cause the groundwater table to rise rapidly and cause shallow buried large diameter pipe to float. Flotation of the pipe can result in failure of pipe sections or over deflection of a joint, providing a path for the introduction of I/I. Deep manholes are also subject to floating in areas with high groundwater tables.

2.1.1.1.4 Blockage resistance

Materials with low roughness constant (e.g., low "*n*" in the Manning Equation) or friction factor will help prevent blockages. The constant range associated with clay pipe is typically 0.0130 to 0.0150, whereas that for fiberglass and polyvinyl chloride (PVC) pipe is 0.0090 to 0.0105. Smooth pipes provide superior flow characteristics and help prevent the buildup of debris that result in blockages, flow restrictions, or reductions in overall system capacity. In addition to pipe material, pipe slope, flow depths, and velocities are also critical and are detailed in Section 2.1.1.3.

2.1.1.1.5 Maintenance

Selected materials should be ones that the operating agency has the proper equipment to maintain. For example, specialized fusion equipment may be required for the repair of a high-density polyethylene (HDPE) pipe failure to the degree of tightness expected if the original piping had fused joints. Alternatively, ductile iron and PVC pipe can be repaired to their original tightness with specialized fittings or by removing and replacing the failed section.

2.1.1.2 Connections

When not completed properly, lateral connections to the main sewer pipes and pipe connections to manholes and other structures can provide an avenue for large quantities of I/I to enter the collection system. These connections can also affect system hydraulics when transitions are rough, involve bends with high angles that change the flow directions, protrude into the main sewer pipe, or provide sites for solids to settle and/or accumulate. Ideally, connections should route flow from that connection tangentially into the main flow stream. Leaky "hammer taps" (knock a hole in the pipe) by plumbers who want to avoid the cost and work involved in installing a proper lateral tap to a main line are common in systems that were developed with inadequate monitoring and controls. Hammer taps also compromise the structural integrity of the pipe, whereas lateral fittings are inherently stronger than the pipes and strengthen the system.

2.1.1.2.1 Minimize infiltration/inflow or sediment intrusion

Minimizing I/I and sediment intrusion into the existing collection system may be the focus of system characterization as discussed in Section 1.1 and require extensive source detection work and subsequent manhole modifications. A sewer use ordinance and a lateral policy may need to be established detailing responsibilities for maintaining laterals, together with requirements for disconnection of roof drains, basement drains, and other sources of inflow. In 2003, *Reducing Peak Rainfall-Derived Infiltration/Inflow Rates—Case Studies and Protocol* (WERF, 2003) it is stated, "The results of the analyses of these projects, supported by the literature survey, strongly indicate that ignoring the private sewers put utilities at risk of not reducing peak RDII flows to any significant degree."

Pipes found with extensive problems may require replacement, slip lining, or repair using techniques such as the cured-in-place-pipe (CIPP) process or pipe bursting. Laterals found with defective connections should be removed and reconnected properly. Consideration should be given to replacement of the lateral to the property or easement line when a defective connection is found. One reference cites in their study that private sector sources contributed 59% of the total I/I flow (Gonwa et al., 2004). Data indicate that repair of defective laterals can reduce peak flows during rain events by 20 to 75%. The WEF Collection Committee maintains a growing library (www.wef.org/PrivateProperty) of case studies from private property-related programs at wastewater utilities that provides further information on technologies, programs, and policies for managing I/I from private sources. Further information on private lateral investigation and repair can be found in *Methods for Cost-Effective Rehabilitation of Private Lateral Sewers* (WERF, 2006).

Pipe connections to new manholes (precast concrete or other construction, such as fiberglass) should be completed using flexible watertight gaskets that are cast into the manhole. The use of grout as the primary sealant to seal pipes in manholes should be avoided because grout does not provide a positive, long-lasting seal. Grout typically will not bond to pipe and will crack as a result of differential settlement, improper curing, or hydrostatic pressure, thereby providing a path for I/I to enter the collection system. Final connections into the manhole should be made with a short section of pipe (approximately 0.6 to 0.9 m, or 2 to 3 ft in length) so that there is a joint close to the manhole. This joint will allow limited differential settlement of the pipe without compromising the system integrity. A typical push-on-type joint will allow deflections of 3 to 5 deg without loss of seal. Mechanical or restrained type fittings are not recommended for use in the connection of piping to manholes.

Existing manholes with multiple defects or significant I/I should be considered for replacement with new manholes of precast concrete or fiberglass construction. Numerous proprietary systems are available for lining of deteriorated manholes and when properly applied will restore the condition of the manhole and reduce I/I. Selection of the rehabilitation approach should consider life-cycle costing results and site factors.

Construction inspection, material testing, and quality control through processes, such as air pressure or vacuum or hydrostatic testing, are critical elements of new construction and rehabilitation of existing facilities to provide for their longevity and eliminate sources of I/I.

Lateral connections are addressed in Section 2.1.2.2.

2.1.1.2.2 Hydraulic transitions

Lateral connections should be completed using only manufactured products with a documented success record. Field connection techniques, such as "hammer tapping", should not be allowed because the loose joints provide sites for the I/I and pipe trench material to enter the pipe and affect the system hydraulics. Options that should be considered for lateral connections are the following:

- In new construction using ductile iron, clay, and PVC sewers, use a wye constructed of the same material with gaskets provided by the pipe manufacturer to provide a watertight seal. For new laterals into existing sewers, use flexible compression-type fittings that will provide a watertight seal. Holes should be cored into the existing pipe using only equipment recommended by the manufacturer to provide a proper fit.
- For fiberglass sewers, use flexible compression-type fittings that will provide a watertight seal. Holes should be cored into the existing pipe using only equipment recommended by the manufacturer to provide a proper fit.
- For CIPP sewers, use full-saddle lateral connections connected to the CIPP liner rather than to the host pipe alone.
- For HDPE sewers, use electro-fusion saddle or for larger diameter connections, use flexible compression-type fittings that will provide a watertight seal. Holes should be cored into the existing pipe using only equipment recommended by the manufacturer to ensure a proper fit.

Manholes should have channels and benches to provide a smooth transition between the influent and effluent pipes and to minimize the buildup of sediment and debris. The channel should extend to the pipe springline if possible. For new manholes, preferences vary from channels and benches, which should be factory grouted, to channels and benches formed in the field. If possible, a drop of 0.03 m (0.1 ft) should be provided across a new manhole to offset hydraulic losses. Sharp changes in pipe directions (greater than 45 deg) should be accommodated using large diameter or multiple smaller diameter manholes to provide a smooth transition.

Existing manholes that were installed without channels and benches or in which the structures have deteriorated should be installed or replaced as required. A section of the effluent pipe should be used to form the channel to create a smooth transition between influent and effluent lines. Proper material selection, installation, and testing are critical for successful field construction.

2.1.1.3 Hydraulics

Collection system hydraulics ultimately determines system capacity. The system capacity can be maximized by the design and operation to include integration between components such as CSO pumping and storage facilities. Regulatory requirements may dictate how components within a system can be used, such as the operation of the CSO facilities and peak flow rates affecting the overall capacity. Requirements may also establish design standards.

2.1.1.3.1 Maximizing flow in system

Improper sizing of collection system components can result in long-term maintenance issues, overflows, and even failures, making redesign and replacement necessary. For example, in a system with piping significantly larger than required for the flows conveyed, crown corrosion resulting from hydrogen sulfide gas released from the slow moving flow may result in a pipe failure. Sizing errors result in unnecessary expenses that can strain the budgets of most service providers. Data collected as part of the preliminary engineering process should provide the basis for properly sizing the conveyance system components.

Rather than embarking on new construction, it is often possible to increase the capacity of existing systems or provide that the maximum capacity is not exceeded through operational and other modifications. Pumping stations discharging into gravity sewers that have capacity issues may be able to use wet wells as wet weather flow equalization basins at the station and induce upstream system storage during wet weather events to reduce or equalize the flow discharged to the gravity sewer. This may be accomplished by modifying pump run times or using a programmable logic controller and flow meter to control the discharge rate. Upstream monitors should be installed to ensure critical elevations are not exceeded. A critical elevation establishes the maximum a system can be surcharged without an overflow or backup of flow into laterals and homes. Installation of backflow devices may be necessary in low areas to help reduce the risk of backup into homes during extreme events. As Section 1.2.3.1 discussed, backflow devices are not fail-safe and can leak, particularly if they do not have an air gap.

Surcharging may provide the pressure (or head) to force additional flow through an existing system, allowing it to convey flow beyond its original design capacity. As discussed above, controls need to be established and critical elevations determined and monitored.

Other options for maximizing flow include diversion of excess flow to areas of the system with remaining capacity. This may be accomplished by surcharging to backup flow into other systems or by installation of additional pipe to convey flow between areas/basins. Conversion of semiautomatic regulators to static regulators may allow additional flow to be forced though the system to the WRRF.

Replacement of piping sections with larger diameter pipe, elimination of negative slope sections, and increasing the slope in sections to the maximum extent allowed by the downstream piping will help eliminate hydraulic restrictions and increase the overall flow velocity and capacity. Criteria for minimum pipe slope by pipe size have been described by numerous documents, including *Recommended Standards for Wastewater Facilities* (GLUMRB, 2004). The documents recommend a minimum velocity of not less than 0.6 m/s (2 ft/sec) when full.

Determination that the capacity of the existing system has been maximized reduces the alternatives to minimization of I/I or installation of a new conveyance system.

2.1.1.3.2 Component integration

As noted in Section 2.1.1.3.1, maximizing the flow through a system may require component integration, such as conveying flow to basins that have remaining capacity. Components to be integrated may include in-line diversion dams and weirs, wastewater pumping stations serving different areas, gravity sewers with pumping stations, and in-line or off-line sewer storage and WRRFs. This integration may be accomplished by elements of real-time control, such as the following:

- Construction and implementation of a comprehensive centralized SCADA system.
- Installation, calibration, and maintenance of depth sensors or permanent flow meters to provide real-time information regarding the flow elevation at critical system locations, or the flow rates of a pumping station or through a section of the collection system;
- Automation of conveyance, storage, and WRRFs to allow remote startup and operation; and
- Maximizing the capacity of the WRRF. This may involve construction of new facilities, modifications to hydraulic limiting structures, or overall process changes.

2.1.1.3.3 Regulatory requirements

Federal, state, regional, local, and other regulatory agencies establish standards by which conveyance systems are designed, constructed, and operated. Many, but not all, state regulatory agencies have established standards for conveyance systems that include flow and design criteria, lateral connection requirements, material of construction, and reference documents, such as *Gravity Sanitary Sewer Design and Construction* (ASCE and WEF, 2007), for items not covered within their regulations.

In states in which these standards have not been established, local plumbing code should dictate the lateral connection requirements and the agency (home/business owner or operating utility) responsible for maintaining the upper and lower laterals. Within the code or other enforceable documents, standards for new conveyance systems should be detailed to include approved material of construction, minimum pipe capacity, sizing, slope, and flow velocity, and installation requirements. Construction inspection and testing must be an integral part of all work to avoid receipt of defective work.

Local agencies should also develop standards for I/I source detection and reduction. These standards should detail procedures to include data collection requirements for manhole inspection, smoke testing, CCTV, and other source detection activities. Drawings and specifications should be prepared that detail the required repair method for each type of defect found and inspection and testing requirements.

Accessibility to collection system piping and manholes should also be addressed as part of these standards. Overflows commonly take place in remote areas in which the system cannot be easily accessed because of the terrain and undergrowth. Many overflows could be prevented with regular inspection and maintenance. For these areas, consider specifying construction of an inexpensive road adjacent to or on top of the line as part of the project so that the manholes and piping can be easily accessed. Gates or other barriers will need to be installed as part of the project to limit access.

Federal and state regulatory agencies establish discharge standards with requirements for the use of combined sewer storage and WRRFs. Options for meeting these standards are discussed in additional detail in the following sections.

2.1.2 *Treatment*

The first section of this chapter presented an overview for strategic wet weather planning. This section provides slightly more depth regarding the factors important in design of WRRFs. Design of a wastewater treatment system capable of handling dry weather flows and peak wet weather flows requires careful consideration and balancing of criteria for system hydraulics and treatment processes. The goal is a WRRF that is cost-effective to construct, operate, and maintain and has sufficient capacity to provide treatment under peak conditions to protect the environment. Several design guidelines are available, including *Design of Municipal Wastewater Treatment Plants* (WEF et al., 2010) with more recent updates to some sections, and *Recommended Standards for Wastewater Facilities* (GLUMRB, 2004) as well as local state guidelines and standards. The locally applicable guidelines provide the criteria that local regulators will use in reviewing and approving facility designs.

For the WRRF, the three areas discussed below—system controls, physical layout of the facilities, and selection and sizing of the unit processes—should be considered during design to ensure that the WRRF will be able to effectively treat peak wet weather flows.

2.1.2.1 Control Systems

Water resource recovery facilities are incorporating more complex control strategies to efficiently treat the variations in wastewater flows and loadings experienced at the WRRF. Including these controls is possible because of improvements in the reliability, accuracy, and cost of field instruments and control devices as well as computer control systems. Controls for WRRF unit processes can range from manual operation at the equipment in the field to remote and fully automatic operation through the WRRF's SCADA system. Details of control strategy development and design of control systems are covered in other reference documents.

As WRRFs become increasingly automated and have higher levels of instrumentation, the operator must rely on the accuracy of that data as decisions are made, especially during critical wet weather periods. To ensure good data, a redundancy and reliability analysis of proposed field instruments and controls must be included in the design process. The accuracy and reliability of the controls should also be periodically assessed after construction and proper calibration and maintenance performed.

2.1.2.1.1 Pumps, valves, and piping

Most activated sludge WRRFs have multiple pumps, valves, and interconnecting pipes to join the treatment unit processes to a treatment system. Considerations in the design of these facilities to enhance the ability to treat wet weather flows will be discussed.

Pumps are used in WRRFs to convey wastewater, sludge, chemicals, and effluent to and from unit processes where gravity flow is not feasible. Pumping systems must be sized with a range of capacity adequate to allow turndown to pump the lowest dry weather flow and enough capacity to pump the peak wet weather flow. This is often satisfied with multiple pumps of varying capacity and may include variable-frequency drives to increase the range of a given pump. Pumps can be started and stopped remotely with the capacity controlled automatically using instrumentation.

Valves and gates are used in WRRFs to isolate equipment and throttle flow. Valves can be controlled remotely to bring equipment into service in response to rising wet weather flows. Valves and gate types should be selected based on the characteristics of the fluid and the type of service (isolation or throttling).

Piping and channels are used in WRRFs to convey wastewater to and from unit processes, in either an open-channel or a pressurized condition. Wastewater can flow by gravity or be pumped through the piping. The designer must take great care in sizing and layout of piping at the WRRF. All of the flow conditions that are expected over the life of the facility must be estimated early in the design, and piping must be provided that will adequately convey the flow.

A key criterion in sizing piping is flow velocity and the type of material that will be conveyed. Wastewater with settleable solids should have a minimum velocity of 0.3 m/s (1 ft/s) with periods of higher velocity to scour any solids that might settle in the pipe at low flows. On the other extreme, high velocities can result in high friction head losses that can limit the peak flow capacity or require high pumping head requirements.

Gravity systems with freeboard available above normal flow conditions can use surcharging of the system to allow higher flows to pass through the system in wet weather conditions. Unit processes, such as primary and secondary clarifiers, can process high flows that submerge the effluent weirs; however, under those conditions, the clarifiers lose the ability to control the flows that can degrade treatment performance.

2.1.2.1.2 Instrumentation

Control of unit processes within a WRRF during wet weather events begins with accurate and reliable data collected and transmitted from field measurement devices. Flow measuring devices, such as flow meters installed in open channels and closed conduits, are easily fitted with local readouts and can transmit instantaneous and historical flow information through the SCADA system. Other field data are easily obtained, such as liquid level in tanks and wet wells using level elements, pump speed from adjustable frequency drive units, pressure in pipes using transducers, valve and gate positions using position indicators, and verification of equipment operation using slow speed switches.

These data can be transmitted via the SCADA system and provide the operator with readily accessible process information throughout the WRRF during a wet weather event. The operator can then make informed decisions to maximize the capacity of the WRRF. Control systems allow operators to bring equipment in and out of service from remote locations. Improved control systems have allowed staffing levels at WRRF to be reduced while maintaining high levels of treatment and process reliability.

2.1.2.2 Unit Layout

A WRRF unit processes can be arranged on the site to improve flexibility for operation and to provide high levels of treatment during wet weather flows. Layout typically begins with determining the ultimate treatment capacity and timeframe envisioned for the WRRF site. That information is typically determined during master and facility planning studies. Once the capacity is determined, WRRF effluent, air emissions, and other performance standards can be estimated throughout the life of the site. Unit processes can be selected and designed to meet estimated performance standards and current standards. It is important to provide for inclusion of future unit processes in laying out the facility and to provide adequate hydraulic head and space between unit processes for future construction.

2.1.2.2.1 Flexibility

An important component in providing effective treatment of wet weather flows is the flexibility available to the operator to bring the appropriate unit process online in a short time. As discussed above, remote and automated startup of individual units can help the operator bring additional units on-line as needed to treat peak flows without requiring numerous operations personnel in the field opening and closing gates and valves.

Most design standards have some additional capacity built into the recommended standards for peak flow conditions. This serves as a safety factor to ensure good performance of a process if the operating flow exceeds the design conditions, as long as it is within that margin of safety.

Unit processes are protected and treatment is improved during wet weather by diverting temporarily portions of the process flow to equalization basins. Flow in the basins is returned to the process flow when system flows allow.

The ability to operate the unit process in both dry and wet weather modes without sacrificing performance is an important element in selecting a unit process. An example of this is the ability to operate an activated sludge system in a conventional mode during dry weather conditions and in step feed or contact stabilization mode during wet weather conditions. Another example of process flexibility is the ability to add a coagulant and flocculant to the influent to a primary clarifier during wet weather conditions to enhance the removal rate of TSS and BOD and increase the surface overflow rate of the clarifier to enhance overall treatment capacity (chemically enhanced primary treatment). A third example is use of a vortex/swirl separator to remove solids and floatables with addition of coagulant and flocculant to enhance removal efficiency. A fourth example of wet weather flexibility that can be designed into the WRRF is in the headworks with influent screening. Using a combination of traditional screens and stormwater screens in the headworks can provide cost-effective screening for average flows and allows for overflow to the stormwater screens during peak flow periods. An example of that screening scenario for headworks is illustrated in Figure 3.16.

2.1.2.2.2 Orientation and shape

Treatment processes should be designed in shapes that maximize uniform splitting of flow to the basins and distribution of flow uniformly across the unit to avoid short-circuiting and maximize performance. Location of the individual basins and reactors on the project site is important to ensure uniform flow split and the ability to add more units to increase capacity in the future. It is important

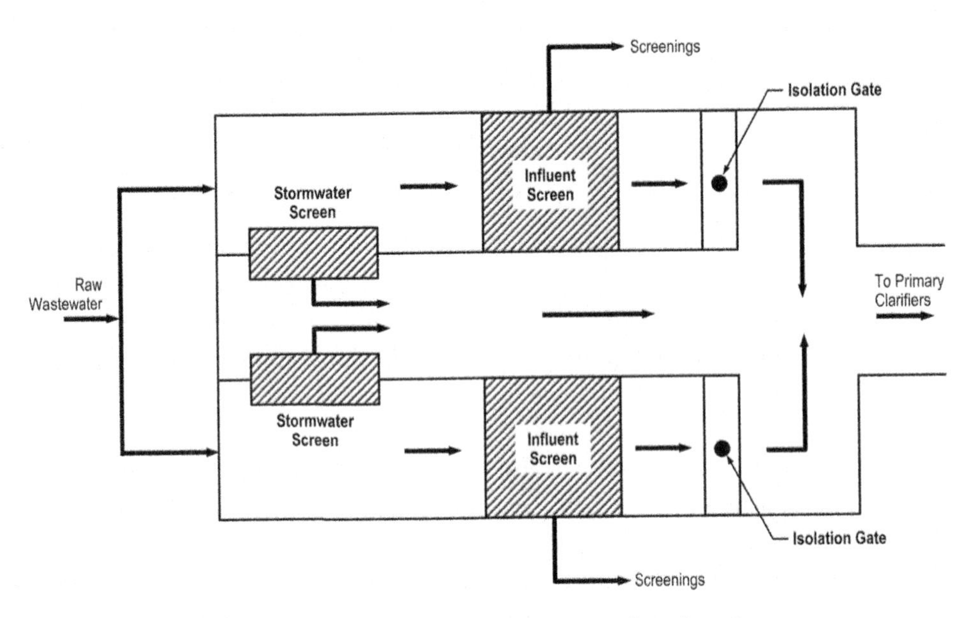

FIGURE 3.16 Incorporating stormwater screens to a headworks.

to arrange facilities to avoid dead ends that would allow solids and floating material to accumulate, restricting flow and creating odor potential.

As discussed below, it is becoming more common to provide piping and channels for activated sludge aeration basins, allowing them to function in the step-feed mode and increasing treatment capacity during wet weather. Step feed can also be successfully incorporated into the design of facilities that remove nutrients (nitrogen and phosphorous) using activated sludge and biological selectors.

2.1.2.3 Process

The type of treatment processes selected for the WRRF can greatly affect the ability of the WRRF to treat wet weather flows. To meet secondary treatment effluent parameters for BOD, a biological process must be incorporated into the overall secondary treatment flow scheme.

2.1.2.3.1 Unit sizing

Treatment processes should be sized and arranged to treat expected low flows and loadings as well as peak flows and loadings. Physical and chemical treatment processes will have set capacities, with wet weather flows accommodated by bringing more units into service.

The rated capacities for physical and chemical treatment processes are set by the manufacturer and by the physical arrangement of the equipment. As with all field installations, the actual capacity will vary depending on the installation details and the wastewater to be treated. To determine the actual capacity of individual units, stress testing under field conditions is an important tool that should be used after installation to verify the performance of the unit processes and confirm that the design capabilities can be achieved.

2.1.2.3.2 Biological systems

Practices exist that can be used in the design of biological treatment systems to maximize treatment capacity where large swings in low to peak flows and loadings are expected. Because biological systems involve the growth of organisms to treat soluble organics, all biological secondary treatment systems will have limits to their capability to treat wet weather flows.

Fixed-film systems have the capability to handle wider fluctuations in short-duration peak flow and loading events than do suspended growth systems such as activated sludge. The biomass is attached to some type of media that has more resistance to being washed out of the system with peak flows. Fixed-film systems typically do not produce as good an effluent as do activated sludge systems and are not used as frequently where low limits of TSS, BOD, and nutrients are required. Hybrid systems that use media for growth of bacteria within activated sludge basins can provide the benefits of both systems (integrated fixed-film activated sludge and trickling filter/solids contact are two examples).

Providing for step feed or contact stabilization feed of wastewater to the aeration basin during wet weather events will enhance secondary effluent by reducing the solids loading in the mixed liquor applied to the secondary clarifiers. When in the step-feed mode, return activated sludge is returned to the front of the aeration basin and primary effluent is split between feed points along the length of the aeration basin, with more of the flow during wet weather sent to the last quarter to third of the basin. This has the effect of reducing the concentration of mixed liquor suspended solids (MLSS) in the portion of the aeration basin that is fed to the secondary clarifiers. The wastewater is brought into contact with the bacteria, which absorb the soluble organics. The bacteria are settled in the secondary clarifier and returned to the aeration basin for further treatment. Step feeding helps to extend the hydraulic capacity of the secondary clarifiers by reducing the solids loading. Step feed for activated sludge is illustrated in Figure 3.17.

Design of secondary clarifiers using current proven technologies can enhance the ability of the clarifiers to settle biological solids during peak flow conditions. Incorporating energy dissipation devices into the inlet of the clarifier can reduce the energy of the mixed liquor and effectively spread the flow uniformly across the clarifier and direct it to the bottom of the clarifier. Devices, such as Stamford and McKinney baffles, are available for use on the periphery of the clarifier to redirect flow back toward the center of the clarifier, further reducing energy currents and enhancing solids separation. Use of these devices has been proven to improve secondary clarifier performance during average and peak flow periods.

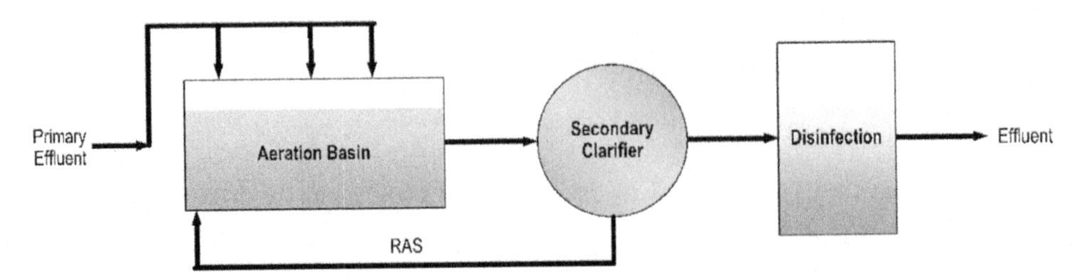

FIGURE 3.17 Step feed for activated sludge (RAS = return activated sludge).

It may also be possible to add a flocculent (polymer) and/or coagulant for a limited period of time to the mixed liquor before it enters the secondary clarifier to enhance the rate at which solids settle in the secondary clarifier, effectively providing additional wet weather treatment capacity. The effectiveness of chemically enhanced flocculation and coagulation in the secondary clarifier should be verified in demonstration testing at the WRRF to determine how much that system can improve capacity. The short duration of chemical addition is to avoid charge reversal of the activated sludge resulting from recirculating the chemically conditioned sludge to the aeration basin.

The designer may include one or more of the tools described above in the design of the facility. The improved wet weather capacity should be verified using field stress testing. All of the techniques are aimed at enhancing the ability of the secondary clarifiers to effectively separate the biological solids from the secondary effluent. This enables the WRRF to produce an effluent that meets permit conditions during wet weather events and returns the biomass to the aeration basin to be available to uptake more soluble BOD, treat the wastewater, and create an environment to grow more biomass.

2.2 Fiscal Management

The cost to implement an LTCP, a proposed CMOM, and other wet weather programs can be significant. A key element for implementation of wet weather controls is the ability to fund the selected controls. For many utilities, wet weather improvements will include relatively costly, capital-intensive projects. The federal CSO Control Policy states that each utility "... is ultimately responsible for aggressively pursuing financial arrangements" for implementation. It is suggested that a financial consultant familiar with municipal finance be part of the planning and/or implementation team.

Financial management for wet weather issues includes both short term and long-term considerations. Short-term financial management is typically fiscal year planning or budgeting. Long-term financial management typically considers a period of 3 to 5 years or longer. For infrastructure assets with relatively long effective life and expensive replacement, there is a greater focus on even longer range planning so that utilities are not financially drained by increasing repair and replacement costs as infrastructure ages.

Ideally, the wet weather improvement projects identified by the utility should be part of the utilities asset management program in which the projects are included in the utilities capital investment validation and prioritization process used for establishing the utilities' long-term rate structure. This process may also include the development of an integrated plan using U.S. EPA's IPF (2012).

Timely investment can result in reduced project costs through efficiencies of scale and avoidance of repeat repair costs. In the long term, cost-effective infrastructure reinvestment will have a positive effect on customer rates.

2.2.1 *Conveyance and Treatment*

A key financial component for any utility is the ability to project the funding needs necessary to responsibly renovate and replace the existing wastewater infrastructure, including both conveyance and treatment. Many of these improvements are driven by the need to handle wet weather flows. A significant portion of this infrastructure was constructed during the post-World War II building booms and is reaching the end of its useful life.

The condition assessment of WRRFs is more straightforward than that of collection systems because WRRFs are visible, accessible, and are in a relatively small, localized area. The assessment of conveyance assets is more difficult because these are underground, extensive in area, and expensive to inspect. Even with a regular program for conveyance condition assessment, there will always be an additional, unidentified component of conveyance that will require repair or replacement resulting from failure or unplanned needs identified during any given year.

Financial planning for conveyance requires the development of an asset management program that considers both an effective condition assessment program, with funding for repair of identified defects, and as a predictive model for the unidentified needs that will surface over time. Because the rate of failure and the needed reinvestment in any given year will not be constant and will tend to echo construction periods from earlier years, it is important that financial planning consider these projections when budgeting system needs.

2.2.2 *Budgeting*

Developing budgets for conveyance and treatment system needs requires regular planning, condition assessment, and review of historical operations and maintenance data. Condition assessment provides the information from which to assess and prioritize repairs that can have a significant effect on system performance, especially during wet weather periods. Historical operating data provide information on locations and facilities in which system performance issues exist and help focus assessment and evaluation efforts.

Several financial forecasting best practices have been developed in the water industry that may have application to wastewater infrastructure. These practices forecast replacement needs over the pipe life cycle.

A variety of infrastructure forecasting tools have been developed resulting from the evolution of asset management programs in the water and wastewater industry. Some of the latest asset management resources and tools can be found through WEF (www.wef.org) and the Water Environment Research Foundation (www.werf.org). Typical average useful life for conveyance facilities is from 50 to 100 years, whereas WRRFs may be from 20 to 30 years. Using asset management tools for determining assets' useful lives, forecast curves for replacement can be developed from which financial needs for replacement can be estimated.

2.2.3 *Costs*

Total budget needs for conveyance systems include O&M, repair and replacement, and capital project costs. Capital projects typically refer to all expenditures that are included in a utility's capital improvement plan or capital investment plan. As previously mentioned, repair and replacement costs can include both identified needs and those that are not identified but are anticipated based on historical trends and/or life-cycle analyses.

Studies conducted by WERF (www.werf.org) and annual surveys conducted by NACWA (www.nacwa.org) offer resources for estimating conveyance and treatment system operating costs.

Improved financial management can be achieved by tracking the costs for performing various activities. A technique commonly referred to as activities based costing (ABC) is a process to quantitatively measure the cost and performance of activities, resources, and cost objects, including overhead when appropriate. The ABC captures organizational costs (including direct and indirect expenses) and applies them to the defined activity structure. The ABC is a process of simplifying and clarifying decisions required by process evaluators and senior management using activity costs rather than gross allocations.

Typical examples of ABC include cleaning costs per linear foot, CCTV costs per linear foot, and repair costs per linear foot. These enable evaluating the cost-effectiveness of service provided by utility staff when compared to outsourcing or other third-party service provider options. Computerized maintenance management systems (CMMS) can facilitate the process of tracking the cost of activities.

2.2.4 *Funding and Affordability*

Funding for conveyance and treatment needs, including renewal and replacement, should be integrated into a utility's system financial planning and funding process. Developing the funding process typically requires assessing alternatives for financial strategies that consider different customer growth rates, changes in debt service, rate increases, and other variables. Key points to include in each financial alternative are the following:

- Funding source assumptions,
- Financial policy assumptions, and
- Operation and maintenance policy and budget assumptions.

These key points may include a review of high- and low-end alternatives and affordability tests. U.S. EPA and other affordability guidelines, such as that by NACWA (NACWA, 2005), are available but do not always provide a definitive result, and the guidelines are not always accepted by regulatory authorities.

Disproportionate effects on disadvantaged populations are a significant reality for many cities implementing wet weather capital plans and compliance programs. Some cities have successfully negotiated financial conditions in their wet weather consent decrees.

3.0 OPERATING AND MAINTAINING

The importance of properly operating and maintaining wastewater systems to prevent or minimize overflows and sewer backups (water-in-basement) has been recognized in NPDES permits, regulatory practice, and guidance. Most dry weather overflows and a significant number of wet weather overflows can be avoided by effective system O&M.

Formal or informal "self-audits" using U.S. EPA's proposed CMOM for separate sewer systems are an effective means of documenting current system conditions and O&M practices as well as identifying gaps where improvements in system operation can be made. Most enforcement actions also require CMOM audits. Many states have detailed O&M requirements for combined systems that include construction observation and oversight.

The standard for evaluating separate system O&M is the *Guide for Evaluating Capacity, Management, Operation, and Maintenance (CMOM) Programs at Sanitary Sewer Collection Systems* (U.S. EPA, 2005a). A similar approach has been developed for combined sewer systems by the New York State Department of Environmental Conservation in its "Best Management Practices Annual Report Checklist" (NYSDEC, 2004). Both of these approaches provide guidance on effective O&M components to minimize and prevent overflows. Additional information regarding the types and frequencies of maintenance activities can be found in *Optimization of Collection System Maintenance Frequencies and System Performance* (ASCE and U.S. EPA, 1999) and *Sanitary Sewer Overflow Solutions* (ASCE and U.S. EPA, 2004), and *Collection Systems: Methods for Evaluating and Improving Performance* (Arbour and Kerri, 1998).

3.1 Monitoring and Reporting

Monitoring and reporting requirements for specific wet weather conditions are now included in most permits and enforcement actions. The requirements are site and parameter specific and intended to be adapted to the conditions and concerns associated with the individual systems. Because CSOs are permitted discharges and SSOs are not, there are differences in how CSOs and SSOs are reported. It is important that each agency adopt procedures so that consistent and reliable data on SSOs and related performance factors can be obtained. A POTW must meet monitoring and reporting requirements in its permits and should evaluate the feasibility and cost of additional monitoring if needed to improve the understanding of system performance. The value of anecdotal information from maintenance staff and others is also important and needs to be "harvested" routinely in an organized approach.

3.1.1 *Conveyance*

At a minimum, monitoring and reporting of overflows from the conveyance system should include information related to the following:

- Location (address and/or asset identification, with a map);
- Receiving water;
- Condition of the site (type of debris, evidence of vandalism, etc.) with photos;
- Duration;
- Estimated volume; and
- Contributing factors and root cause of the overflow.

This information is useful for documenting system performance and to determine what, if any, additional action should be taken. Additional action could include increased maintenance activities, preparations for special procedures during wet events, studies, or system capital improvements. Reporting requirements to regulatory agencies often include "oral reports" within 24 hours by telephone, fax, or e-mail for events that may affect public health or are greater than an established minimum. Follow-up written reports are required within a specified time period, typically within 5 days. Frequent and recurring CSOs that have consistent and predictable effects may be subject to periodic summary reporting, rather than reporting on individual events. Commonly required information includes

- Cause of the overflow;
- Steps taken to reduce, eliminate, and prevent reoccurrence of the overflow;
- Schedule to reduce, eliminate, and prevent reoccurrence of the overflow; and
- Steps taken to mitigate the effects of the overflow.

Requirements that are even more rigorous may apply in some circumstances if receiving water sensitivity warrants:

- Sampling to assess effects and
- Special notice to affected persons, including third parties and, most frequently, state or local health departments and sometimes including public notice through media advisories or the Internet.

Record keeping for a specified time (at least 3 years) and annual summary reports may be required.

3.1.1.1 Performance Monitoring

The frequency and magnitude of SSOs or CSOs are measures of system performance that can be used to evaluate effectiveness of system O&M. Conveyance system performance monitoring technologies range from very low-tech to high-tech methods. It is important that the performance-monitoring program include a thorough understanding of system hydraulics and likely weak points in the system during peak flow conditions. These likely weak points can be identified by a review of system record drawings, flow monitoring, hydraulic modeling, or a combination thereof. Performance monitoring can be enhanced by field checks during peak conditions and through educational programs to obtain feedback on overflows from the public. The level and type of system performance monitoring should be structured to cost-effectively meet both the service needs of system customers and regulatory requirements.

Monitoring performance will prove useful for system assessment and improvement alternatives evaluation. Robust data recording and documentation systems will enhance the future value of the data.

3.1.1.1.1 Tools and processes

Many different tools and processes are used to monitor wet weather system flow and to detect overflow conditions. The simplest technologies may be no more complicated than drawing chalk marks or placing blocks of wood on weirs to detect if an overflow has occurred. In sensitive or hard-to-reach locations, real-time monitors may be used to detect and measure overflow activity. Selection of the right technology is site specific and typically designed to meet both permit requirements and system operating needs. Temporary flow monitoring to develop models and assess wet weather strategies is a common practice. Many utilities have their own flow monitors that they use constantly to assess system performance. Many larger systems deploy permanent real-time monitoring at key flow locations to provide more complete and timely information about system performance and to detect and investigate abnormal flows.

Modeling is also used as a means to identify or report overflows as an alternative to difficult and expensive (if not impossible) site-specific monitoring. The system model is run for the specific storm event, and overflows predicted to have occurred by the model are reported whether or not they were observed or detected in the field. Of course, the models must first be calibrated and validated to confirm that calculations and predictions are representative of actual conditions and acceptable to satisfy regulatory scrutiny.

In both large and small systems, overflows can be assessed and reported based on observations of the system maintenance personnel using their experience and whatever guidance they may have been given. Similarly wet weather flows in many collection systems are based on observed surcharging or overflows and flows to the WRRF, not on sophisticated monitoring. Many overflows occur in conditions that cannot be detected or measured by conventional means, especially when systemwide flooding is occurring. In these situations, reports should still be submitted if there is evidence that an overflow is occurring with an estimate of the overflow volume using best available information.

3.1.1.1.2 Criteria

Criteria for the use of specific performance monitoring technologies for conveyance systems have not been established. Regardless, it is important that the performance methods used to gather data are useful for evaluating the operation of the system, improve the understanding of the level of service being provided, and are selected to be commensurate with the magnitude of problems that may exist. The performance-monitoring program should be structured in a cost-effective manner to protect the environment, public health, and to meet system-operating requirements.

Regulatory guidelines do not identify specific technologies that must be used and how often they should be used, but regulatory agencies expect that each utility will understand the state of its system and will report overflows properly. System ignorance and lack of knowledge does not relieve a utility from this requirement. Significant enforcement actions have been taken for no other reason than inadequate or inaccurate reporting of overflows.

Systems with CSOs must meet the nine minimum controls (NMC) and demonstrate compliance. Specific performance standards or technologies are not required, but a utility must be proactive in assessing and implementing O&M practices that can help reduce or eliminate CSOs.

3.1.1.2 Reporting and Notification

Reporting and notification must be conducted to meet the specific requirements of typical permitting authority regulations and guidance, system operating requirements to protect customers and prevent system damage, as well as any additional requirements imposed by regulatory orders. Failure to report violations that are known or observed can be considered civil or criminal infractions with potentially serious penalties or consequences to the utility, and possibly responsible individuals. Environmental management system (EMS) requirements focus heavily on monitoring and reporting and implementation of an EMS can assist a utility in tracking and reporting regulatory information more consistently. Ideally, reporting and notification should be integrated into the utility's information management systems, enabling the decision-makers to evaluate results and respond through improved operations, maintenance, or other improvements if the reports indicate trends in causes or locations.

The importance of a utility fully complying with reporting requirements cannot be understated. Concerns about public health effects are driving wet weather policy, and immediate reporting may be expected with no tolerance for delays. For dry weather overflows, reporting and immediate mitigation are expected. Utilities should be aware that there are sometimes different reporting requirements or protocols for wet versus dry weather overflows.

3.1.1.2.1 Information dissemination

Information dissemination requirements typically are not always identified in permits or enforcement agreements, although specificity in this area is becoming more common. A standard is emerging that encourages "transparency" in wet weather monitoring and reporting. Transparency means that the information is available to all interested parties concurrently without special procedures such as Freedom of Information Act requests. No standards exist and system operators should include wet weather reporting protocols in their communication strategies, making sure to meet minimum regulatory requirements and expectations of affected and interested parties. Using the Internet and agency Web sites to disseminate information is becoming the most cost-effective and accepted approach. For situations in which this approach will not provide adequate communication with affected portions of the community, supplements such as public meetings should be considered. Internal system operating information needs must have a high priority, especially for extreme event response and protection of customers and property from backups and system failures.

3.1.2 Treatment

Monitoring and reporting requirements for POTWs in dry weather, normal flow conditions are well established and understood. Specific criteria for sampling locations, flow monitoring, parameters, testing frequency, and reporting are included in every NPDES permit (although, even with national guidance, there are significant differences between requirements of specific permitting authorities and special conditions that may relate to receiving waters or other conditions).

Numeric criteria for conventional pollutants typically are based on 1-, 7-, and 30-day averages. These averages may mask variations associated with high flows. Nevertheless, if high flows cause exceedances of permit criteria, the permittee may be subject to enforcement action associated with the permit violation. Some permitting authorities issue permits that have special language for wet weather conditions, and the frequency of special wet weather monitoring and reporting requirements at POTWs is increasing.

3.1.2.1 *Performance Monitoring*

Treatment system monitoring and reporting is a real-time, 24 hours per day/ 7 days per week activity that follows well-established protocols and procedures. Information is collected and reported via discharge monitoring reports to permitting authorities and must meet recognized standards for sampling and testing. Flow monitors are part of the WRRF infrastructure that must be maintained and operated properly. Permit requirements specify what parameters and flows are to be monitored and reported and the frequency and occurrence of these activities.

Although treatment performance monitoring is well established, there are variations in what is required to be monitored during wet weather events that may have a dramatic effect on WRRFs. Treatment units and even the entire WRRF may have to bypass part of the flows to avoid damage to process units or flooding the WRRF. System operators must react quickly to changing conditions, and performance monitoring may not be perceived as a priority during these events. Operators should be knowledgeable regarding the reporting requirements even under these circumstances.

To assess treatment system performance in wet weather and optimize treatment strategies to allow maximum treatment of wet weather flows, more information is needed than is typically required in NPDES permits. Special monitoring and evaluation of WRRF performance in wet weather flow conditions are needed and may be required in some circumstances. In facilities with storage, equalization, or parallel wet weather treatment facilities, more information may be needed for operators to respond to changing flows. In some cases, additional automated flow monitoring equipment may be needed to manage flows, to meet NPDES reporting requirements, or to document bypass conditions.

3.1.2.1.1 Tools and processes

Stress testing of wet weather process capability is a highly recommended practice. To develop wet weather *standard operating procedures* (SOPs) that are meaningful to operators, the reaction of process units to high flows must be determined through process performance testing. Optimizing treatment efficacy and avoiding washout of solids or essential biomass is also a consideration that can affect WRRF performance for days or weeks after a significant rain event.

3.1.2.1.2 Criteria

Criteria for monitoring and reporting at WRRFs are well established and included in federal and state regulation and guidance.

3.1.2.2 Reporting and Notification

Reporting and notification requirements for WRRFs are included in most NPDES permits. In many jurisdictions, spills, emergency WRRF bypasses, or failures that could affect receiving waters must be reported immediately, as noted above, often with a more detailed follow-up report. In some situations, this could include performance variations caused by wet weather flows. Notification of others, especially health departments, may be required in some states or regions. The NPDES reporting requirements have been subject to continual evaluation. Modifications to NPDES permit requirements may result in additional monitoring and reporting requirements. With the current regulatory emphasis on wet weather, special attention should be given to draft permits that may include additional monitoring and reporting.

3.1.2.2.1 Information dissemination

The NPDES monitoring and reports are public information and available to all interested parties. Proactive utilities make this information easily available and may include special notice of overflows or other incidents that may affect performance. Use of the Internet and Web site reporting is becoming common, especially in larger utilities.

3.2 Performance Assurance

Even if not required by a permit, special attention must be given to operating and maintaining equipment and standard operating procedures to ensure optimum performance in wet weather flow conditions when some or all of the conveyance system and WRRF may be stressed by high flows. Documentation of wet weather O&M performance assurance methods is becoming a standard requirement through U.S. EPA's CMOM guidelines and other regulatory requirements.

3.2.1 Conveyance

Conveyance system performance assurance should focus on preventing or reducing backups and overflows. Performance measures must be developed and implemented with considerable attention to documentation and consistency in data gathering and interpretation. The regulatory standard for conveyance system operation is that "the permittee must properly manage, operate, and maintain at all times, all parts of the collection system that the permittee owns or over which it has operational control". A good reference that describes effective management and tools for maintaining conveyance systems is *Core Attributes of Effectively Managed Wastewater Collection Systems* (APWA et al., 2010).

Mitigation of the effects of overflows is also an O&M expectation. The recommended approach includes the following key performance assurance elements (among others):

- Provide access to necessary locations and have the ability to undertake necessary actions for appropriate emergency response,
- Implement the general and specific prohibitions of the national pretreatment program under 40 CFR 403.5,
- Control fats and grease and system blockages from roots and/or situations in which capacity is being reduced by accumulations of solids,
- Reduce overflows caused by power outages and equipment breakdowns,
- Develop and implement an overflow emergency response plan, and
- Conduct sufficient monitoring and evaluation of system performance as needed to determine the effectiveness of O&M procedures in place.

Guide for Evaluating Capacity, Management, Operation, and Maintenance (CMOM) Programs at Sanitary Sewer Collection Systems (U.S. EPA, 2005a) contains a checklist of 149 questions concerning most aspects of conveyance system O&M as well as utility finance and management. The checklist provides a method to assess current operating practices and system performance. Although there is no national SSO policy, U.S. EPA recommends and supports CMOM approaches, and some enforcement actions include required CMOM audits. Figure 2.4 shows where the CMOM Program is integrated into a utility's performance implementation cycle.

A CMOM approach does not provide for definitive steps on how to fill any identified data gaps or system deficiencies. Many questions relate to issues that may appear to have no or little linkage to SSOs, such as operator safety, customer complaints, spare parts inventory, and public relations. It is up to each utility and the regulators to decide if the practices of a utility are causative factors in the frequency and severity of SSOs or other performance-related problems.

Wide-ranging and expensive changes in utility O&M programs can result from a CMOM audit. Wastewater utilities are well advised to self-audit or use an independent auditor to assess their CMOM practices and identify gaps or potential liabilities.

3.2.1.1 *Observations and Decisions*

Basic decisions about conveyance system performance begin with two questions:

- Do you have frequent and recurring SSOs and/or CSOs?
- Do you have frequent and recurring customer service issues such as sewer backup problems?

If the answer is yes, there will be an expectation of any utility from a regulatory standpoint to eliminate or reduce the SSOs and to address CSOs through an LTCP, and from a costumer standpoint, to reduce the incidence of sewer backups.

Although O&M upgrades may be less expensive in the short term and quicker to implement than LTCPs and capital projects to eliminate overflows,

O&M upgrades typically affect operating budgets and priorities and may require fundamental management changes in how the utility is operated, especially with respect to performance assurance and documentation.

3.2.1.1.1 Tools and practices

The O&M tools and practices that address all of U.S. EPA's proposed CMOM and NMC regulatory expectations include a long list of items, most of which are fundamental to daily utility operations and require no special considerations for wet weather flow management. A few tools and practices have a special significance in conveyance system management and O&M when applied to wet weather compliance. These include the following:

- System mapping, characterization, and use of GIS;
- System monitoring and reporting (including the use of SCADA as appropriate);
- Sewer ordinance enforcement;
- Computerized maintenance and management information systems for SSO and work order tracking; and
- Sewer inspection, cleaning, repair, rehabilitation, and/or replacement.

3.2.1.1.2 Criteria

There are no performance evaluation criteria applicable to wet weather that can be considered absolute. Anything that is known or suspected of contributing to excessive SSOs or CSOs is subject to evaluation and could be included in a Section 308 letter (regulatory information request) from U.S. EPA, which typically is a first step leading to enforcement action and a wet weather compliance inspection. Section 308 letters typically require POTWs to provide extensive documentation and records.

At a minimum, system operators should know the following:

- Maximum and average flows;
- Procedures for finding and measuring overflows;
- Annual number of overflows;
- Causes of overflows (e.g., blockage, pump malfunction, overloaded sewer, construction damage);
- Locations and causes of sewer collapses;
- Grease program statistics and information about blockages and overflows from fats, oils, and grease (FOG);
- Statistics on sewer cleaning and inspection; and
- Pumping station inspection frequency.

3.2.1.1.3 Analysis

Utilities are expected to assess information collected and reported. Applications, such as GIS and CMMS, can make this type of monitoring/reporting and assessing more efficient and effective. It is especially important that utilities have an analysis of their system performance in wet weather, often referred to as a system characterization. Response plans, including inspection and cleaning schedules, should be a derivative of system analysis and characterization and the assessment of monitoring and reporting information.

3.2.1.1.4 Prioritization

Priorities for O&M and capital improvements must be established that address performance failures leading to overflows or other system problems. To establish priorities effectively requires a good understanding of system performance, system hydraulics, and environmental effects. Prioritization methods using risk-based decisions are discussed in detail in Chapter 2.

3.2.1.1.5 Operations and maintenance practice adjustments

It is important that utilities regularly review their O&M practices to optimize, to the best of their ability, their O&M program. Optimization of O&M can be achieved by reviewing the types, frequencies, and effectiveness of various practices. The relationship between O&M practices and system performance is a key consideration when making O&M practice adjustments.

3.2.1.2 Sewer Cleaning and Debris Removal

Utilities should immediately address SSOs and CSOs (especially dry weather overflows) that result from debris, sediments, or other obstructions, because this is not only a relatively easy fix but is required to regain original system capacity. In addition, obstructions causing overflows will be given special attention by regulators.

3.2.1.2.1 Scheduling criteria

Frequency of system O&M practices that could affect overflows and system performance are site specific and must be determined locally based on experience considering generally accepted industry practices. Nevertheless, through reported practices and research, there is a general acceptance of reasonable and effective types and frequencies of many activities. Utilities must provide that adequate resources are available to implement the selected O&M plan. Because utilities have limited budgets, scheduling O&M should be optimized and responsive to system-specific conditions. Although less frequent inspection and cleaning activities may result in overflows in some systems, some practices being performed may not be effective and do not produce results that justify the expense. Operators must base their needs on industry practices and site-specific experience. Every utility should document efforts to assess system performance and set priorities.

3.2.1.2.2 Scheduling techniques

Use of CMMS and prioritization methods that use modeling and other computerized approaches are gaining traction in conveyance system management and O&M. Field applications of GIS and computers using remote reporting and access to data are becoming common.

3.2.1.3 Emergency Response Plans

It is expected that system O&M include emergency response planning that will allow responses to situations that are producing overflows. These plans should include response protocols and mitigation of damage, but also specify reporting requirements and follow-up activities that may prevent recurrence of the problem.

3.2.1.3.1 Power outages

Power outages and response to power outages have been identified as a significant area of concern to regulators. Overflows caused by power outages are not acceptable, and utilities are expected to have standby power or generators that will allow the system to continue to operate without overflows.

3.2.1.3.2 Overflow and spill response planning

Response planning is an important component of overflow prevention and mitigation. Utilities are expected to respond immediately to overflows and spills, and take actions to eliminate the overflow and mitigate any damage to the environment. Documentation of actions taken and reporting is also a component of good response planning. Security measures related to vandalism or public health and safety should also be considered.

3.3 Construction

Construction is typically not considered as predominant a factor influencing wet weather as other "minor processes" discussed in this guide; however, construction activities can significantly affect wet weather, particularly if the design is not executed or built as intended. The construction "minor process" is not just a responsibility of the contractor but also of the owner. The design intent should be clearly presented in the design documents, and it is imperative that the owner or owner's construction representative be diligent about keeping the design intent at the forefront of daily construction activities.

Construction is an extension of the design. Although the contractor's work objectives may include satisfying the owner, completing the project as quickly as possible at the least cost is typically the contractor's highest priority. The owner's highest priority is for the constructed system to perform as planned and designed. A well-designed system and documented project construction management and construction observation are the construction mechanisms available to the owner in a traditional design, bid, and construction project. Oversight of construction activities by individual(s) with the appropriate level of experience and expertise greatly increase the chances of the system being installed and performing as designed.

3.3.1 *Conveyance*

Conveyance systems, by their very nature, reveal many unknowns during construction. These unknowns can include varying subsurface conditions such as demolition debris, rock, high/varying groundwater tables, utilities, contaminated soils, or even unstable/unsuitable soils. Designers must decide the balance between incorporating the required structural strength in the pipe material or the field-installed support bedding.

The design process must include a site/route investigation to confirm that the project is feasible and to help reduce the potential for change orders during construction. However, it is challenging to provide that adequate subsurface data have been collected and unknowns eliminated resulting from the cost and typical challenges associated with field mapping. Utility owners often do not maintain accurate record drawings and utilities are often missed during field mapping. Issues encountered during construction resulting from unknowns can and will result in field adjustment. Without field adjustments, the design life or design intent of a conveyance component may be significantly shortened or changed, respectively. For instance, soils encountered during installation that have varying support strength require field changes to adjust pipe bedding, compaction method, and/or backfill materials. Without these field adjustments, pipe joints might separate or deflect beyond acceptable tolerances and allow extraneous water to enter the system. Deflections could also occur midspan in pipe sections, reducing the capacity of the system because of formation of air pockets during peak events or the buildup of heavy solids and/or causing damage to the internal linings that are provided for corrosion protection.

Without construction observation and testing, the owner's planned objective cannot be confirmed under controlled conditions. Even with construction observation and testing, the performance of the constructed facility may change unexpectedly immediately or relatively soon after construction. Presented below are practices that help reduce the occurrence of extraneous water and help ensure that the system performs as intended.

3.3.1.1 *Construction Observation*

Construction observation and testing go hand in hand. They are used by both the contractor and the owner to confirm that the intent of the design was achieved. Owners share different inspection philosophies. Some place primary inspection confidence in performance tests that judge the acceptance of the construction work at the end of the project or activity. Little, if any interim construction observation is performed. Another philosophy is to watch the construction activities on a continuous or regular basis and conduct confirming tests at the end of the project or activity.

The first philosophy of construction observation can produce more wet weather maintenance or rehabilitation requirements sooner in the components' expected life span than the latter construction observation philosophy. Typically, the construction activity that can result in maintenance or rehabilitation is covered up and cannot be confirmed without considerable disruption and conflict. The issues result from activities performed during construction that have workmanship errors, or activities that fall short of their requirements for the constructed system to perform as intended. Example activities prone to accelerate the degradation of the component include improper surface cleaning for material adhesion; the substitution or use of inferior, inappropriate or defective materials; or not temporarily sealing the ends of constructed pipelines, thus allowing rains to wash sediment into the pipe and possibly cause accelerated erosion of pump volutes and impellers within a downstream pumping station. Continuous construction observation and testing is a preferred practice to reduce the risk of wet weather conveyance problems.

The following are several common construction observations that are performed to confirm that the design intent is achieved:

- Closed circuit televisions for rehabilitated and new pipes and manholes: CCTV will confirm the presence of construction debris or sediment, active leaks, and possibly pipe sags. One practice is to run water in the pipe immediately before the CCTV to better note any ponding. Ponding can confirm pipe sag or a pipe laid flat or at reverse grade. Any of these deficiencies will reduce the design capacity of the pipe and become hydraulic restrictions in the system.
- Perform grade surveys of the installed force main pipe to identify the highest elevation in the designated run of pipe so that the air valve can allow the maximum amount of air entrainment in the pipe to be released. Air trapped in a main can substantially reduce capacity and provide a site for corrosion.
- Confirm that trench excavation and backfill materials are appropriately applied to meet the pipe bedding and backfill requirements. This includes making sure the trench is properly dewatered and the appropriate bedding and backfill materials are used. Improper pipe bedding is a leading cause for conveyance systems to deviate from their expected capacity performance and can result in sags within the pipe, separation of joints, and damage to pipe corrosion protective lining system.

3.3.1.1.1 Testing

Testing is another mechanism to determine adherence to the design intent and the potential for a constructed system component to become degraded. Tests that have proven effective are the following:

- Vacuum tests for manhole rehabilitation and new construction, which tests improve workmanship and reduce I/I. Vacuum tests appear to provide a more assuring and definitive test compared to a hydrostatic test.
- Spark testing and holiday testing of liners in manholes and pipelines will help confirm that the coating system was uniformly applied and that there are no holes that can provide sites for the start of corrosion that could cause premature failure.
- Mandrel test for new, nonrigid pipe materials is preferred toward the end of the project, or at least a month after installation, to allow any bedding or backfill deficiencies to reveal themselves in the form of excessive deflection. Deformed pipe has reduced capacity and can become sources of I/I.
- Low-pressure air tests for new gravity pipe installations confirm that pipe joints and pipe connections are watertight.
- Hydrostatic pressure tests of new force main pipe installations confirm that pipe joints, connections, and valves are watertight.

3.3.1.1.2 Contractor practices

Owners can provide contractors with an incentive toward good workmanship to reduce construction deficiencies that eventually lead to wet weather problems. One simple but effective method is to clearly state in the contract documents that the owner will pay for construction observation and testing conducted by the owner throughout the project. If owner construction observations or testing reveals that the contractor's work is not meeting the specified requirements, the contractor will reimburse the owner for the extra construction observation and testing. The owner should specify that the owner can increase the frequency and number of tests if desired and that the reimbursement will continue until a specific sequence or number of construction observations or tests show the work is again acceptable.

New sewers are often installed by private developers who construct the conveyance system according to the utility's requirements and then turn over ownership of the system to the utility. Several utilities have discovered service lateral problems, particularly during wet weather, that have passed the air test inspection. The source of these issues is commonly related to the increased use of trenchless installation of other utilities, such as gas and cable that result in lateral pipes being penetrated. To reduce this occurrence, private developers are required to sequence their utility construction so that the service lateral is the last utility installed.

3.3.1.2 *Site Work*

Designers and construction managers should consider the beneficial construction activities the contractors will inherently incorporate into their work to execute the project, and expand on them to improve the operations and maintenance of the constructed project. Several examples are discussed below regarding rights-of-way access and runoff control.

3.3.1.2.1 Rights-of-way access

Conveyance owners must consider access to remotely located conveyance sewers and force mains a key component of the conveyance system plan to facilitate the sewer's O&M. Contractors include in their construction bids provisions for clearing rights-of-way and building roads for getting their equipment to a majority of the project site. Therefore, incorporating in the design an enhanced road bed or surface topping for the owner's use at the end of the project can help resolve an age-old access problem at an economical value. Typically, many wet weather problems—both infiltration and exfiltration—can be resolved when there is access to inspect and repair large sewers located near streams and in river flood plains.

3.3.1.2.2 Best management practices plan and final site grading

All construction projections should include a prepared best management practices (BMPs) plan for sediment and erosion control. For most pipeline projects, a construction stormwater permit will likely be required by the regulatory agency. The requirements vary often by location. The plan should be fully implemented and updated to include additional BMPs as required to preclude material from leaving the project site and to protect the installed work.

Controlling surface water runoff during construction may not seem like an immediate threat or have an apparent link to wet weather issues, but it does. Surface grading and installed BMPs should avoid leaving a manhole or partially installed pipe in an area susceptible to ponding. Even if the manhole materials are initially constructed relatively watertight, the manhole can develop defects from external conditions that become sources for I/I. Flooding of an area can affect pipe bedding and backfill resulting in long-term settlement or deformation issues.

Pipelines built along steep embankments are subject to unbalanced embedment forces and can experience separated joints or broken pipe if surface water erosion exposes segments of the buried pipeline. Stone or other erosion-resistant covering, appropriate vegetation, and maintenance will help keep the pipe or manhole stabilized.

3.3.2 *Treatment*

As with the conveyance system, construction of WRRFs is an extension of the design. The constructed facilities should provide the capacities for treatment that were intended during the facility planning and design process. On-site inspections should be performed to ensure the constructed facilities match the design documents, and performance testing should be done during startup to ensure that the facilities perform as intended.

3.3.2.1 *Construction Observation*

Observation during construction includes review of equipment shop drawings, construction materials, subsurface conditions, foundations and supports, elevations of structures and weirs, and subsurface structure backfill. It is important to confirm that all elements of the intent of the facility design are incorporated into the constructed facilities. Proper observation by the owner, designer, or a third-party construction manager can provide that the constructed facilities are true to the design.

Proper installation of piping, pumps, valves, and gates can provide that the wet weather capacity and flexibility of the facility is provided as designed. Setting piping and flow control structures and devices at the intended elevations is crucial to maintaining the design capacity.

Providing appropriate materials of construction for items, such as anchor bolts, connectors, piping, coatings, and structural elements, is important for extending the life of the facilities. Water resource recovery facilities typically require corrosion-resistant materials that are a significant upgrade over conventional construction materials. Care must be exercised during submittal review and during construction to confirm that the materials included in construction are the quality of the materials specified.

3.3.2.2 *Testing*

After construction is completed, there is typically a period of functional testing and performance testing of the facilities. It is during this time that the owner verifies that the installed facilities will perform as intended during the design. The construction contract documents will typically specify the conditions under which the facilities will be tested and the performance that is required. It is important to verify the wet weather treatment capacity of the facilities during this testing period.

Testing the facilities under wet weather conditions can be difficult to achieve, especially if the return frequency of wet weather events is small. The low probability of a wet weather event occurring during the performance-testing period can be a challenge to certifying the equipment. An approach to solving this dilemma is to isolate individual units, direct all of the flow to that unit, and observe its performance. There are reference documents that can help the operator establish the necessary protocol to plan, execute, and document this stress testing.

3.3.3 *Treatment*

It is to be expected that WRRF operators develop response strategies based on experience and have response plans to deal with wet weather flows. Typically, these strategies are intended to maximize treatment of wet weather flows and reduce WRRF bypasses, prevent damage to WRRF facilities or processes from high flows, and ensure permit compliance. At times, and under severe stress, these goals can be conflicting. Wet weather SOPs for WRRFs must anticipate flow conditions, identify critical process locations and control points that must be adjusted when flow conditions warrant (which may be before the wet weather flow actually reaches the facility), ensure that critical capacity to handle wet weather events optimally is online, and otherwise provide guidance to operators dealing with rapidly changing and highly variable flow conditions.

3.3.3.1 *Observations and Decisions*

Operators at WRRFs are constantly observing and evaluating flows and system performance. Before the emphasis on wet weather compliance, operators were mainly concerned with preventing damage to facilities and maintaining process integrity when experiencing high volume wet weather flows. Most WRRFs had been (and many still are) designed to handle wet weather flows 1.5 to 3 times dry weather flows at capacity, typically determined by U.S. EPA or state guidelines that set a limit. These high flow values often were proven inadequate, leaving WRRF operators to fend for themselves when higher flows exceeding process capacity were experienced. Operators were (and are) forced to bypass excessive flows or modify their treatment strategies based on observation and experience to provide the maximum treatment to flows and achieve permit compliance whenever possible.

It is now more common that WRRFs are designed with (or are being required to provide) wet weather capacity and facilities to handle high flows with greater certainty and performance outcomes. Operator observations and decisions are more important than ever in achieving the best performance results managing wet weather flows. Management strategies described in this guide can assist operators in developing strategies to maximize performance and minimize risks during wet weather events and in documenting performance to support regulatory compliance.

3.3.3.1.1 Tools and practices

Operators of large facilities typically have assigned process control personnel who use modeling and assessment tools to optimize system performance in all conditions and now with a special emphasis on managing wet weather flows. Long-term control plans and SSO abatement plans are including options that send more flow to the WRRF for longer periods of time.

Smaller systems may or may not have access to the same tools or expertise but can apply what they learn from observation and operating experience under a variety of wet weather flow conditions that can be documented in wet weather SOPs.

3.3.3.1.2 Parameters

Permit conditions and standards drive performance assurance at WRRFs. Because most permits do not address wet weather flows, operators are left on their own to force fit their normal operating and monitoring to wet weather conditions and permit monitoring and reporting. Sometimes this is easily accomplished, but often there are changes occurring in high volume flows that affect process performance, including limitations on secondary treatment systems to effectively treat wet weather flows and achieve the performance requirements specified in most permits. Other permit conditions may be met in high flow conditions. Mass balance in systems with CSOs is typically meaningless in that permits do not identify surface loadings during wet weather flows. Utilities with increasing wet weather performance expectations at their WRRFs should work with permit authorities and provide operators with as much guidance and support for managing wet weather flows as is possible.

3.3.3.1.3 Analysis

Analysis of WRRF performance under the dynamic circumstances of wet weather conditions is essential. Water resource recovery facilities should be stress-tested under controlled conditions and process operating ranges and limits for high-volume wet weather flows need to be clearly identified. Information on stress testing is available in *Secondary Settling Tanks: Theory, Modeling, Design and Operation* (Ekama et al., 1997) and *Upgrading Wastewater Treatment Plants* (Daigger and Buttz, 1998). Models are available to evaluate the capacity of the WRRF to hydraulically process high flows and treatment models are available from simple static spreadsheet models to complex dynamic simulation models to evaluate physical, chemical, and biological treatment processes performance.

Routine settling tests, sludge depths in clarifiers, MLSS concentration analysis, mean cell residence time, waste activated sludge wasting rates, sludge volume index, and return activated sludge flow rates are available to provide a measure of health of the WRRF and its ability to process wet weather flows.

3.3.3.1.4 Practice adjustments

Adjustments can be made before the start of and during wet weather events to maximize treatment capacity and performance. The WRRF operators must be aware of the predicted precipitation and use that to anticipate flow conditions in the conveyance system, including the intensity and duration of projected peak flows so they can be prepared for those flows when they arrive at the WRRF. Weather predictions are available on-line from the National Weather Service or through several commercial media outlets. There are also weather prediction services that will provide subscribers with on-line predictions, weather alert pages, and frequent e-mail updates throughout the duration of a storm. As discussed in Section 1.2.3, Evaluate Alternatives to Optimize Wet Weather Flows, numerous operational adjustments are available that can be used in response to predicted and observed weather conditions to improve capacity and performance of WRRFs during wet weather events. The following discussion focuses on related practice adjustments for activated sludge-based WRRFs that should be considered for implementation by operators either before or during the occurrence of the actual event:

- Minimizing down time for all critical equipment at the WRRF during the wet weather season; scheduling routine maintenance during the dry season;
- Maximizing the number of secondary clarifiers in service;
- Creating an environment in the bioreactors that encourages formation of good settling biomass;
- Minimizing the inventory of biomass retained in the secondary clarifiers by maintaining a low sludge blanket;
- Utilizing step feed in the bioreactors to reduce the biomass loading to the clarifiers;
- Adding, for a short period of time, polymer and/or coagulant to the biomass at the end of the aeration basin to enhance settling in the secondary clarifiers; and
- Turning off mixers/aeration, for a short time, to selected bioreactors to settle and store biomass in the basins and reduce the solids loading to the secondary clarifiers. This concept is illustrated in Figure 3.18.

As mentioned earlier in this document, stress testing of the unit processes in the WRRF should be performed under high flow conditions to better understand the capacities of each of the elements in the WRRF and to identify hydraulic and treatment performance bottlenecks before the start of wet weather events. Guidelines for stress testing under controlled field conditions are available. Compiling the results of the individual unit stress testing into an overall picture of expected WRRF performance will provide the utility with a prioritized list of necessary improvements to increase the WRRF's overall wet weather treatment capacity. The results will also provide guidance to the operator for proper operation of the facility before and during the wet weather events.

3.3.3.2 Control Alternatives

Depending on the treatment processes designed into the WRRF and permitted operating scheme, the operator has several control alternatives available to increase wet weather treatment capacity while maintaining a high-quality

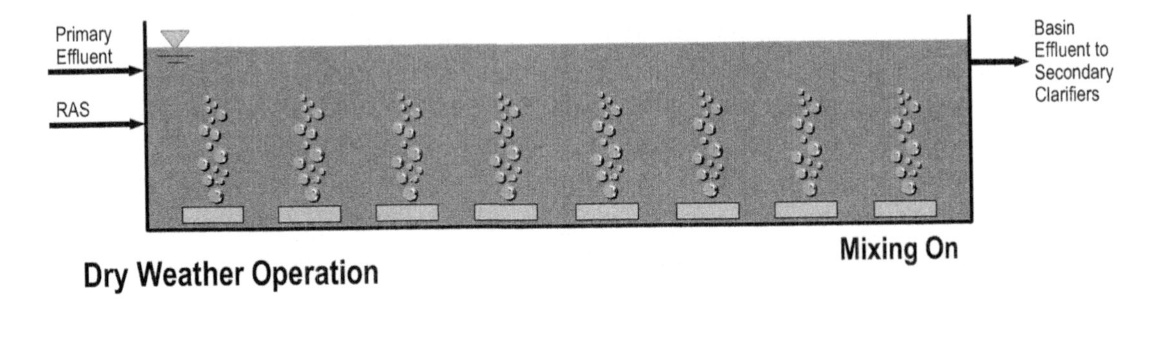

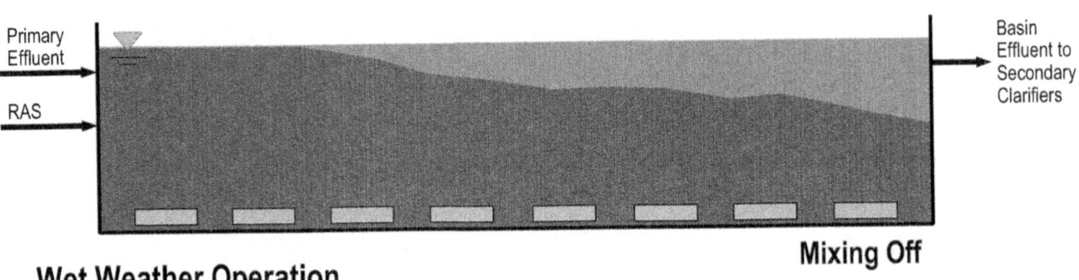

FIGURE 3.18 Solids storage in an aeration basin during wet weather (RAS = return activated sludge).

effluent. Several of these alternatives were discussed in detail in Section 1.2.4.4, Treatment, and are summarized below. As stated there, the availability and applicability of the alternatives ultimately depends on regulatory agency and other stakeholder requirements for the WRRF. It should also be noted that incorporating these additional control capabilities at a WRRF will increase the overall operational and maintenance requirements, thus requiring careful consideration of the necessary additional resources required in terms of staff, training, and budgets.

- On-site storage: Reduces the effect of the extreme flow and load conditions associated with wet weather events; results in reduction of facility requirements; and provides the operator with additional time to adjust and manage appropriately the event.
- Maximizing and protecting existing capacity: Provides operators with facilities in which hydraulic bottlenecks have been eliminated and with the means to temporary enhance and protect process performance (such as chemical enhancement of primary clarifiers and step-feed conversion for activates sludge processes), and gives them the tools needed to increase capacity while maintaining long-term processing capabilities.
- In-facility flow rerouting: Protects physical and processing capabilities of existing systems; maximizes the use of existing assets; provides the

operator with means to increases on a temporary basis the treatment capacity of a WRRF; and

- Incorporating additional treatment: Several technologies are available that can provide the operator with the additional capacity needed to handle wet weather conditions in those circumstances in which the prior three alternatives are not sufficient to handle the anticipated extent of the design event. Processes discussed in previous sections include fine and micro-screening, vortex/swirl solids separation, chemically enhanced and high-rate clarification, high-rate filtration, and supplemental disinfection.

4.0 REFERENCES

American Public Works Association; American Society of Civil Engineers; National Association of Clean Water Agencies; Water Environment Federation (2010) *Core Attributes of Effectively Managed Wastewater Collection Systems;* National Association of Clean Water Agencies: Washington, D.C.

American Society of Civil Engineers; U.S. Environmental Protection Agency (1999) *Optimization of Collection System Maintenance Frequencies and System Performance;* Prepared by Black & Veatch Corporation for the American Society of Civil Engineers under a Cooperative Agreement with the U.S. Environmental Protection Agency, Office of Water; U.S. Environmental Protection Agency: Washington, D.C.; Feb.

American Society of Civil Engineers; U.S. Environmental Protection Agency (2004) *Sanitary Sewer Overflow Solutions;* Prepared by Black & Veatch Corporation for the American Society of Civil Engineers under a Cooperative Agreement with the U.S. Environmental Protection Agency, Office of Wastewater Management; U.S. Environmental Protection Agency: Washington, D.C.; April.

American Society of Civil Engineers; Water Environment Federation (2007) *Gravity Sanitary Sewer Design and Construction,* 2nd ed.; ASCE Manual and Reports on Engineering Practice No. 60; WEF Manual of Practice No. FD-5; American Society of Civil Engineers: New York.

Arbour, R.; Kerri, K. (1998) *Collection Systems: Methods for Evaluating and Improving Performance;* Prepared for the U.S. Environmental Protection Agency Office of Wastewater Management by the Sacramento Foundation; California State University: Sacramento, California.

Bennett, D., Rowe, R.; Strum, M.; Wood, D.; Schultz, N.; Roach, K.; Adderley, V. (1999) *Using Flow Prediction Technologies to Control Sanitary Sewer Overflows;* Water Environment Research Foundation, Project No. 97-CTS-8; Water Environment Research Foundation: Alexandria, Virginia.

Code of Federal Regulations. EPA Administered Permit Programs: The National Pollutant Discharge Elimination System, Title 40 Chapter 1 Subchapter D Part 122.

Coello Coello, A.; Lamont, G.; Van Veldhuizen, D. (2007) *Evolutionary Algorithms for Solving Multi-Objective Problems*, 2nd ed.; Springer Science + Business Media, LLC: New York.

Daigger, G.; Buttz, J. (1998) *Upgrading Wastewater Treatment Plants*, 2nd ed.; Water Quality Management Library—Volume 2; Technomic Publishing Company: Lancaster, Pennsylvania.

District of Columbia Water and Sewer Authority (2002) *Combined Sewer System Long Term Control Plan Final Report;* Prepared by Greeley and Hansen, LLC: Indianapolis, Indiana; July.

Ekama, G. A.; Barnard, J. L.; Gunthert, F. W.; Krebs, P.; McCorquodale, J. A.; Parker, D. S.; Wahlberg, E. J. (1997) *Secondary Settling Tanks: Theory, Modeling, Design and Operation;* International Association on Water Quality Scientific and Technical Report No. 6; IWA Publishing: London.

Foy, B.; Sands, K. (2004) Public Involvement and the Watershed Approach. *Proceedings of the Collection Systems 2004 Water Environment Federation Specialty Conference: Innovative Approaches to Collection Systems Management* [CD-ROM]; Milwaukee, Wisconsin, Aug 8–11; Water Environment Federation: Alexandria, Virginia.

Gonwa, W.; Simmons, T.; Schultz, N. (2004) Development of Milwaukee MSD's Private Property Infiltration and Inflow Control Program. *Proceedings of the Collection Systems 2004 Water Environment Federation Specialty Conference: Innovative Approaches to Collection Systems Management* [CD-ROM]; Milwaukee, Wisconsin, Aug 8–11; Water Environment Federation: Alexandria, Virginia.

Great Lakes–Upper Mississippi River Board of State and Provincial Public Health and Environmental Managers (2004) *Recommended Standards for Wastewater Facilities;* Health Education Services: Albany, New York.

Metropolitan Sewer District of Greater Cincinnati (2006) *Wet Weather Improvement Plan;* Metropolitan Sewer District of Greater Cincinnati: Cincinnati, Ohio; June.

Milwaukee Metropolitan Sewerage District (2007) *2020 Facilities Plan, Appendix 1A, Production Theory;* Milwaukee Metropolitan Sewerage District: Milwaukee, Wisconsin.

National Association of Clean Water Agencies (2003) *AMSA Wet Weather Survey, Final Report;* National Association of Clean Water Agencies: Washington, D.C.; May.

National Association of Clean Water Agencies (2005) Financial Capability and Affordability in Wet Weather Negotiations; White Paper; National Association of Clean Water Agencies: Washington, D.C.

New York State Department of Environmental Conservation (2004) Best Management Practices Annual Report Checklist; Division of Water, Albany, New York. http://www.dec.state.ny.us/website/dow/csobmp.pdf (accessed May 2006).

Schultz, N.; Munsey, F.; Parente, M. (2004) Deep Tunnels for Sewage Storage and Conveyance: Lessons and Opportunities. *Proceedings of the 77th Annual Water Environment Federation Technical Exhibition and Conference* [CD-ROM]; New Orleans, Louisiana, Oct 2–6; Water Environment Federation: Alexandria, Virginia.

Slaper, T. F.; Hall, T. J. (2011) The Triple Bottom Line: What Is It and How Does It Work? *Indiana Bus. Rev.*, **86** (1).

U.S. Environmental Protection Agency (1994) *Combined Sewer Overflow (CSO) Control Policy;* EPA-830/B-94-001, Office of Water; U.S. Environmental Protection Agency: Washington, D.C.; April.

U.S. Environmental Protection Agency (1995a) *Combined Sewer Overflows—Guidance for Long-Term Control Plan;* EPA-832/B-95-002, Office of Water; U.S. Environmental Protection Agency: Washington, D.C.; Sept.

U.S. Environmental Protection Agency (1995b) *Combined Sewer Overflows—Guidance for Nine Minimum Controls;* EPA-832/B-95-003, Office of Water; U.S. Environmental Protection Agency: Washington, D.C.; May.

U.S. Environmental Protection Agency (1999) *Combined Sewer Overflows—Guidance for Monitoring and Modeling;* EPA-832/B-99-002, Office of Water; U.S. Environmental Protection Agency: Washington, D.C.; Jan.

U.S. Environmental Protection Agency (2004) *Report to Congress: Impacts and Control of CSOs and SSOs;* EPA-83/R-04-001, Office of Water; U.S. Environmental Protection Agency: Washington, D.C.; Aug.

U.S. Environmental Protection Agency (2005a) *Guide for Evaluating Capacity, Management, Operation, and Maintenance (CMOM) Programs at Sanitary Sewer Collection Systems;* EPA-305/B-05-002, Office of Enforcement and Compliance Assurance; U.S. Environmental Protection Agency: Washington, D.C.; Jan.

U.S. Environmental Protection Agency (2005b) National Pollutant Discharge Elimination System (NPDES) Permit Requirements for Peak Wet Weather Discharges From Publicly Owned Treatment Works Treatment Plants Serving Separate Sanitary Sewer Collection Systems. *Fed. Regist.*, **70** (245), 76013.

U.S. Environmental Protection Agency (2005c) *Proposed EPA Policy on Permit Requirements for Peak Wet Weather Discharges from Wastewater Treatment Plants Serving Sanitary Sewer Collection Systems. Fed. Regist.*, **70** (245), 76013.

U.S. Environmental Protection Agency (2010) *Innovative Internal Camera Inspection and Data Management for Effective Condition Assessment of Collection Systems;* EPA-600/R-10-082; Office of Research and Development, National Risk

Management Research Laboratory, Water Supply and Water Resources Division; U.S. Environmental Protection Agency: Cincinnati, Ohio.

U.S. Environmental Protection Agency (2012) Integrated Municipal Stormwater and Wastewater Planning Approach Framework; Office of Water and Office of Enforcement and Compliance Assurance; U.S. Environmental Protection Agency: Washington, D.C.; May.

Water Environment Federation (2006) *Clarifier Design*, 2nd ed.; Manual of Practice No. FD-8; Water Environment Federation: Alexandria, Virginia.

Water Environment Federation (2007) *Wastewater System Capacity Sizing Using a Risk Management Approach;* Technical Practice Update; Water Environment Federation: Alexandria, Virginia.

Water Environment Federation (2010) *Wastewater Collection Systems Management*, 6th ed.; Manual of Practice No. 7; Water Environment Federation: Alexandria, Virginia.

Water Environment Federation (2011) *Prevention and Control of Sewer System Overflows*, 3rd ed.; Manual of Practice No. FD-17; Water Environment Federation: Alexandria, Virginia.

Water Environment Federation; American Society of Civil Engineers; Environmental & Water Resources Institute (2010) *Design of Municipal Wastewater Treatment Plants,* 5th ed.; WEF Manual of Practice No. 8; ASCE Manuals and Reports on Engineering Practice No. 76; Water Environment Federation: Alexandria, Virginia.

Water Environment Research Foundation (2002) *Best Practices for Treatment of Wet Weather Wastewater Flows*; WERF 00-CTS-6; Water Environment Research Foundation: Alexandria, Virginia.

Water Environment Research Foundation (2003) *Reducing Peak Rainfall-Derived Infiltration/Inflow Rates—Case Studies and Protocol;* Project No. 99-WWF-8; Water Environment Research Foundation: Alexandria, Virginia.

Water Environment Research Foundation (2006) *Methods for Cost-Effective Rehabilitation of Private Lateral Sewers;* 02-CTS-5; Water Environment Research Foundation: Alexandria, Virginia.

Glossary

Avoidable Legal term of art meaning that a consequence could have been prevented with the exercise of reasonable engineering judgment in facilities planning/implementation and/or adequate management, operations, and maintenance practices.

Bacteria Microscopic, unicellular organisms, some of which are pathogenic and can cause infection and disease in animals and humans. Most often, nonpathogenic bacteria, such as fecal coliform and enterococci, are used to indicate the likely presence of disease-causing, fecalborne microbial pathogens.

Biochemical oxygen demand (BOD) A measurement of the amount of oxygen utilized by the decomposition of organic material, over a specified time period (typically 5 days) in a wastewater sample; it is used as a measurement of the readily decomposable organic content of a wastewater.

Biological treatment system That portion of a treatment system which is based on biotic activity.

Blending The practice of diverting a part of peak wet weather flows at water resource recovery facilities (WRRFs) around the secondary treatment process units, which are typically biological treatment units, and combining effluent from all processes before discharge from a permitted outfall. Although WRRFs sometimes divert around tertiary facilities, the "blending" discussion has focused specifically on secondary treatment, because of the Clean Water Act technology requirement.

Bypass The intentional diversion of wastestreams from any portion of a treatment facility [40 CFR 122.42(m)].

Capacity The design maximum flow, or loading, that a wastewater system and its components can handle in a specified period of time with predictable and consistent performance. Also, the peak flow is equal to the maximum flow only when the time periods are the same.

Chemical oxygen demand (COD) A measure of the oxygen-consuming capacity of inorganic and organic matter present in wastewater. COD is expressed as the amount of oxygen consumed in mg/L. Results do not necessarily correlate to the biochemical oxygen demand because the chemical oxidant may react with substances that bacteria do not stabilize.

Chemical treatment Any water or wastewater treatment process involving the addition of chemicals to obtain a desired result, such as precipitation, coagulation, flocculation, sludge conditioning, disinfection, or odor control.

Clean Water Act (CWA) The act passed by the U.S. Congress to control water pollution. Formally, the Federal Water Pollution Control Act (or Amendments) of 1972 and all subsequent amendments [(P.L. 92-500), 33 U.S.C. 1251 et seq., as amended by: P.L. 96-483; P.L. 97-117; P.L. 95-217, 97-440, and 100-04].

Coliform bacteria Rod-shaped bacteria living in the intestines of humans and other warm-blooded animals.

Collection system Conveyance system, including intercepting sewers, sewers, pipes, pumping stations, and other structures that convey liquid waste for treatment. Collection system applies to separate sanitary and combined sewer systems.

Combined sewer overflow (CSO) A discharge of untreated wastewater from a combined sewer system at a point before the headworks of a water resource recovery facility.

Combined sewer system (CSS) A wastewater collection system owned by a municipality (as defined by Section 502(4) of the Clean Water Act) that conveys domestic, commercial, and industrial wastewater and stormwater runoff through a single pipe system to a water resource recovery facility.

Controls Processes and/or activities that contribute to removing pollutants from wastewater or to containing and conveying wastewater for treatment and discharge.

Conventional pollutants As defined by the Clean Water Act, conventional pollutants include biochemical oxygen demand, total suspended solids, fecal coliform, pH, and oil and grease.

Conveyance system Often used interchangeably with "collection system", but more broadly can include stormwater, aqueducts, and other modes of transporting liquids.

Cryptosporidium A protozoan parasite that can live in the intestines of humans and animals. *Cryptosporidium parva* is a species of Cryptosporidium known to be infective to humans.

Designated uses Those uses specified in state or tribal water quality standards (WQS) regulations for a particular segment, whether or not they are being attained [40 CFR 131.3(g)]. Uses so designated in WQS are not meant to specify those activities or processes that the waterbody is currently able to fully support. Rather, they are the uses/processes that the state or tribe wishes the waterbody to be clean enough to support, whether or not the waterbody can, in its current conditions, fully support them.

Disinfection The selective destruction of disease-causing microbes through the application of chemicals or energy.

Escherichia coli (E. coli) Coliform bacteria of fecal origin used as an *indicator organism* in the determination of wastewater pollution.

Enterococci Streptococcus bacteria of fecal origin used as an *indicator organism* in the determination of wastewater pollution.

Extreme wet weather Storms that are the high end of the range in terms of rain/snow amounts.

Feasible alternatives The legal term of art used in the "bypass" regulation to identify alternative controls that are both technically achievable and affordable [40 CFR 122.42(m)].

Fecal coliform Coliforms present in the feces of warm-blooded animals.

Flow equalization Transient storage of wastewater for release to a sewer system or treatment process at a controlled rate to provide a reasonably uniform flow.

Giardia lamblia A protozoan parasite responsible for giardiasis. Giardiasis is a gastrointestinal disease caused by ingestion of waterborne *Giardia lamblia*, often resulting from the activities of beavers, muskrats, or other warm-blooded animals in surface water used as a potable water source.

High-rate clarification (chemically enhanced clarification) A physical–chemical treatment process used to treat wet weather flows. Coagulants and/or flocculants are added to the wastewater before entering a gravity settling process. The process creates conditions under which dense flocs with a high settling velocity are formed, allowing them to be removed efficiently at high surface overflow rates with corresponding high total suspended solids and biochemical oxygen demand removal.

Hydrograph Chart or mechanism for recording data that shows variation in flow with time.

Impaired waters A waterbody that does not meet criteria to protect the designated use(s) as specified in the water quality standard.

Indicator bacteria (indicator organisms) Bacteria that are common in human waste. Indicator bacteria are not harmful in themselves but their presence is used to indicate the likely presence of disease-causing, fecalborne microbial pathogens that are more difficult to detect. Coliform, fecal coliform, *E. coli*, and *Enterococcus* are all used as indicator bacteria in water quality standards. Although bacteria have typically been used as indicators of human fecal contamination, other organisms may be used.

Inflow Water that enters the sanitary sewer system directly via depressed manhole lids and frames, downspouts, sump pumps, foundation drains, areaway drains, and cross-connections with storm sewers.

Infiltration Groundwater, including any rainfall runoff that filters through the soil, that enters the collection system through leaking pipes, pipe joints, and manhole walls.

Legal term of art Word that has a particular meaning when used in legal matters, such as discussion of regulations or statutes.

Level of service A quantitative measure of performance for water resource recovery facility functions, which is used to specify the quality of service provided.

Optimum management of wet weather flows To make systems and facilities as functional and effective as possible. Site and process specific.

Pathogen As used in the industry, organisms capable of causing disease, including disease-causing bacteria, protozoa, and viruses.

Performance The manner or efficiency in which an activity or process functions, operates, or is accomplished to fulfill a purpose. Performance has goals or objectives against which measures are established to make corrections to ensure what is about to happen is what was intended and that it conforms to plan, such as reducing pollutants from wastewater to a targeted amount.

Physical treatment A water or wastewater treatment process that uses only physical methods, such as filtration or sedimentation.

Physical–chemical treatment Nonbiological treatment processes that use a combination of physical and chemical treatment methods.

Pollutograph Chart or mechanism for recording data that shows variation in a pollutant's load with time.

Practices Assembly or grouping of protocols.

Preliminary treatment Treatment steps including comminution, screening, grit removal, preaeration, and/or flow equalization that prepare wastewater influent for further treatment.

Primary treatment First steps in wastewater treatment wherein screens and sedimentation tanks are used to remove most materials that float or will settle.

Principles Criteria, engineering rules, and other underlying facts and factors used to evaluate and select controls to achieve the municipality's goals and objectives.

Processes/activities Facilities, treatment systems, management activities, and other activities/actions to contain and convey wastewater and treat that wastewater to remove pollutants.

Protocol A sequenced set of controls (activities and processes) prescribed for a particular purpose that implements the principles.

Publicly owned treatment works (POTW) As used in this publication, refers to the municipality or agency, as defined in CWA §502(4), which has jurisdiction over the treatment works (as defined by Section 212 of the Clean Water Act) and its operation [40 CFR 403.3(o), CWA §212, CWA §502(4)].

Reasonable engineering judgment As a legal term of art, this is the statutory and regulatory standard for evaluating engineering practices.

Regulation Rules, which have the force of law, issued by an administrative or executive agency of government, typically to enact the provisions of a statute.

Rainfall-derived infiltration/inflow The portion of a sewer flow hydrograph above the typical dry weather flow pattern.

Root cause The most basic cause or a problem, which if fixed, will prevent or reduce the likelihood of a problem's recurrence.

Sanitary sewer overflow (SSO) An untreated or partially treated wastewater release from a sanitary sewer system.

Sanitary sewer system (SSS) A municipal wastewater collection system that conveys domestic, commercial, and industrial wastewater, and limited amounts of infiltrated groundwater and stormwater, to a water resource recovery facility. Areas served by sanitary sewer systems often have a municipal separate storm sewer system to collect and convey runoff from rainfall and snowmelt.

Satellite sewer system Combined or separate sewer systems that convey flow to a water resource recovery facility owned and operated by a separate entity.

Secondary treatment Technology-based requirements for direct discharging municipal water resource recovery facilities. Standard is based on a combination of physical and biological processes for the treatment of pollutants in municipal wastewater. Standards are expressed as a minimum level of effluent quality in terms of: 5-day biochemical oxygen demand, suspended solids, and pH (except as provided for special considerations and treatment equivalent to *secondary treatment*) [CWA §301(b) and 40 CFR 125.3].

Sewerage Facilities that provide transport or treatment of sanitary sewage. This guide uses the term "collection system" instead of sewerage.

Stakeholders Parties with a direct interest, involvement, and investment, or that are likely affected by wet weather-related effects and management programs. *Stakeholders* can include the public; environmental interest groups; municipal, county, state, and federal governments; recreation resource management agencies; public health agencies; landowners; point source dischargers; ratepayers; local, state, regional, and federal regulators; state and federal fish and wildlife agencies; and others.

Statute A written law passed by a legislative body of the government.

Technology-based effluent limit Effluent limitations applicable to direct and indirect sources, which are developed on a category-by-category basis using statutory factors, not including water quality effects [CWA §301(b)].

Total suspended solids (TSS) A measure of the filterable solids present in a sample (of wastewater effluent or other liquid), as determined by the method specified in 40 CFR Part 136.

Treatment works Defined in Section 212 of the CWA as "any devices and systems used in the storage, treatment, recycling, and reclamation of municipal sewage or industrial wastes of a liquid nature to implement section 201 of this act, or necessary to recycle or reuse water at the most economical cost over the estimated life of the works, including intercepting sewers, outfall sewers, sewage collection systems, pumping, power, and other equipment, and their appurtenances; extensions, improvements, remodeling, additions, and alterations thereof; elements essential to provide a reliable recycled supply such as standby treatment units and clear well facilities; and any works, including site acquisition of the land that will be an integral part of the treatment process (including land use for the storage of treated wastewater in land treatment systems prior to land application) or is used for ultimate disposal of residues resulting from such treatment" (CWA §212).

Upset An exceptional incident in which there is unintentional and temporary noncompliance with technology-based permit effluent limitations because of factors beyond the reasonable control of the permittee. An *upset* does not include noncompliance to the extent caused by operational error, improperly designed treatment facilities, inadequate treatment facilities, lack of preventive maintenance, or careless or improper operation [40 CFR 122.42(n)].

U.S. EPA guidance document A memorandum or other document written by U.S. EPA to provide information or advice on a regulatory matter. A guidance document may or may not have formal legal status depending on factors, such as the level of agency review, opportunity for public comment, or use of the guidance in compliance and enforcement actions.

U.S. EPA policy Guidance that has been subject to a formal review process and published in the *Federal Register*, for example, the CSO Control Policy.

Virus Smallest biological structure capable of reproduction; infects its host, producing disease.

Wastewater Sanitary and combined wastewater.

Wastewater system Used in this document to mean all municipal wastewater collection and treatment facilities that make up the water resource recovery facility. For planning purposes when there are satellite sewer systems, the satellite sewer systems may also be included.

Wastewater resource recovery facility (WRRF) Used in the industry interchangeably with wastewater treatment plant, wastewater reclamation facilities, and publicly owned treatment works, to refer to the facilities owned, operated, and/or controlled by the municipal or agency "owner"; designed to provide treatment (including recycling and reclamation) of municipal wastewater and industrial waste.

Water quality Legal term of art intended to describe the level of pollutants or other constraints within receiving waters.

Water quality standard A law or regulation that consists of the beneficial use or uses of a waterbody, the numeric and narrative water quality criteria that are necessary to protect the use or uses of that particular waterbody, and an antidegradation statement (40 CFR 130.3).

Water quality-based effluent limitations Effluent limitations applied to dischargers when technology-based limitations are insufficient to result in the attainment of water quality standards (CWA §302).

Waters of the United States All waters that are currently used, were used in the past, or may be susceptible to use in interstate or foreign commerce, including all waters subject to the ebb and flow of the tide. *Waters of the United States* include but are not limited to all interstate waters and intrastate lakes, rivers, streams (including intermittent streams), mudflats, sand flats, wetlands, sloughs, prairie potholes, wet meadows, playa lakes, or natural ponds (see 40 CFR §122.2 for the complete definition).

Wet weather event A discharge from a combined or sanitary sewer system that occurs in direct response to rainfall or snowmelt.

Wet weather flows Additional flows in wastewater conveyance and treatment system, and stormwater, because of weather.

Index

CPSIA information can be obtained
at www.ICGtesting.com
Printed in the USA
BVOW11s0537190816
459274BV00004B/15/P

9 781572 782785